SpringerBriefs in Computer Science

Series Editors
Stan Zdonik
Peng Ning
Shashi Shekhar
Jonathan Katz
Xindong Wu
Lakhmi C. Jain
David Padua
Xuemin (Sherman) Shen
Borko Furht
V.S. Subrahmanian
Martial Hebert
Katsushi Ikeuchi
Bruno Siciliano

T0214020

For further volumes:
http://www.springer.com/series/10028

Tom H. Luan • Xuemin (Sherman) Shen • Fan Bai

Enabling Content Distribution in Vehicular Ad Hoc Networks

 Springer

Tom H. Luan
School of Information Technology
Deakin University
Burwood, VIC, Australia

Fan Bai
ECI Lab
General Motors Corporation
Warren, MI, USA

Xuemin (Sherman) Shen
Department of Electronic
  and Computer Engineering
University of Waterloo
Waterloo, ON, Canada

ISSN 2191-5768   ISSN 2191-5776 (electronic)
ISBN 978-1-4939-0690-1   ISBN 978-1-4939-0691-8 (eBook)
DOI 10.1007/978-1-4939-0691-8
Springer New York Heidelberg Dordrecht London

Library of Congress Control Number: 2014933945

Printed on acid-free paper

Springer is part of Springer Science+Business Media (www.springer.com)

Preface

Our lives nowadays are more absorbed by a variety of content distribution services, like live and on-demand video streaming on Youtube and Netflix, bulk file download through BitTorrent and social information distribution over Facebook. In the meantime, vehicles and Internet also become very prominent parts of our lives. As a direct result, it can be envisioned that the must-have option for vehicles would be the ultra high-speed broadband connectivity that can provide the vehicle travelers the same exciting media content distribution services on the road as they have at home.

Delivering the high-quality live and on-demand contents to fast-motion vehicles, however, is challenged by the unique nature of vehicular communications. On one hand, vehicles have fast mobility which results in the dramatically changing channel conditions and accordingly the turbulent download performance of vehicles. On the other hand, media streaming and content distribution services typically have very stringent Quality-of-Service (QoS) requirements. Therefore, how to fully explore the opportunistic communication connectivity of vehicles and provide the users with their desired service quality is the key. In this monograph, we try to provide an overview of the issue and discuss on three proposals with each one targeting on one specific aspect of this area. To be specific, Chap. 1 introduces the vehicular communication network and the design criteria and challenges content distribution in vehicular networks. Chaps. 2 and 3 consider the on-demand distribution of medium-size content files in the infrastructure-based and infrastructure-less vehicular networks, respectively. In Chap. 2, we describe the proposal to form a large-scale cost-effective infrastructure to enable QoS guaranteed city-wide content distribution services. As building and maintaining a large-scale infrastructure are not only costly but also technically challenging, the presented proposal can address above concerns by building a fully distributed infrastructure with the shared investment cost and fully distributed operating method. In Chap. 3, an integrity-oriented content transmission framework is described. Note that content transmission over the infrastructure-less inter-vehicle connections suffers from the frequent interruptions and may terminate before the content is entirely transmitted. The partially transmitted contents can hardly be successfully presented by the upper-layer applications. The transmission of such contents not only delays the response time of end host users, but it also wastes the precious bandwidth of vehicular networks to deliver useless data. The integrity-oriented content transmission framework targets on the issue with a fully distributed algorithm. Chapter 4 discusses on the distribution of large-volume contents through the live streaming pattern. With vehicles subscribing to the live streaming services but downloading with dramatically changing throughput, this chapter describes an adaptive media playout method which can adaptively determine the playback strategy of users towards the smooth media presentation in the dynamic vehicular networks. Chapter 5 draws a conclusion on the discussed issues and envisions the future research directions.

We would like to thank Mr. Ning Lu, Nan Cheng, Ms. Miao Wang from Broadband Communications Research Group (BBCR) at the University of Waterloo, Dr. Lin X. Cai from Huawei Technology, Dr. Jimin Chen from Zhejiang University for their valuable comments and contributions to the presented research works. Special thanks are also due to the staff at Springer Science+Business Media: Susan Lagerstrom-Fife and Courtney Clark, for their help throughout the publication preparation process.

Burwood, VIC, Australia   Tom H. Luan
Waterloo, ON, Canada   Xuemin (Sherman) Shen
Warren, MI, USA   Fan Bai

Contents

Chapter 1
Introduction

1.1 Vehicular Communications

The advent of advanced communication technologies in the past decades has ushered the era of information age. To date, the high-rate instant access to the media-rich information (e.g., online gaming and video on demands) and ubiquitous communications to each other (e.g., emails and social networks) have already become an integral part of our daily lives. As reported in [1], the average Canadian spends 18 h a week on the Internet, which has outpaced the time spent on watching TV.

Although vehicles nowadays represent the third place, after home and office, where a regular citizen spends more time daily, the high-rate and ubiquitous information access and communications in vehicles are still not available or very expensive in most regions of the world. As reported in [2], Americans spend 15 h on average in a car each week. This is to say, in around 8.9 % of a day, people would have very limited or no Internet access at all. The surge demand of ubiquitous access thus drives the vehicular communication ever more important.

Catering to the ever-increasing demand of vehicular communications, the vehicular networks (alternatively known as vehicular ad hoc networks (VANETs)) have recently been proposed. By equipping vehicles with the on-board wireless transceivers and computerized control modules, and meanwhile deploying the Internet gateways on the roadside (namely roadside unit (RSU)), the vehicles are able to communicate and connected as a network in two modes: the inter-vehicle communication (alternatively known as vehicle-to-vehicle (V2V) communication) for data message exchange among peer vehicles, and the vehicle-to-RSU (V2R) communication to provide the Internet access to vehicles with the assistance of RSU.

Over such a platform, three thrusts of application services can be delivered to vehicles on the move, from the aspects of road safety, driving efficiency and comfort, respectively. The first thrust is related to the road safety. By exchanging and propagating the road traffic information among vehicles using the V2V communications, drivers can be effectively notified on the traffic conditions and accordingly

T.H. Luan et al., *Enabling Content Distribution in Vehicular Ad Hoc Networks*,
SpringerBriefs in Computer Science, DOI 10.1007/978-1-4939-0691-8_1,
© The Author(s) 2014

respond timely to avoid the possible accidents and other extreme consequences. The second thrust is as a key enabling technology of the Intelligent Transportation Systems (ITS). For example, using the vehicular networks, real-time road traffic information can be delivered to vehicles, and guide drivers to optimize their driving routes to avoid traffic jams and reduce the travel time. The third thrust is on the infotainment applications as a means to enhance the travel comfort of passengers. For example, through the V2V communications in proximity, passengers on the board can transfer files among each other, chat online or play games on the road. By connecting to the Internet through road communications infrastructure, i.e., RSUs, travellers can surf the web, watch live TV program or online videos and access media-rich Internet services just as they have at home. Thanks to above applications, vehicular networks can make our trips more pleasant, efficient and much safer than ever before.

In this monograph, we mainly focus on the third thrust. In particular, note that all the digital information needs to be converted into certain form of content files for storage and transmission, efficient content transmission and distribution represent the fundamental and foremost issues of any on-top applications in general and vehicular infotainment services in specific, such as live video streaming, on-demand video/audio service and social communications. As motivated above, this monograph focuses on the discussion of content distribution issues in vehicular networks. We unfold our journey by describing the enabling technologies, recent research developments and proposals towards the efficient, resilient and scalable content distribution to vehicles through both infrastructure-based and infrastructure-less vehicular networks.

1.1.1 Enabling Technology of Vehicular Networks

1.1.1.1 On-Board Equipment

Since the first steam-powered vehicle was invented for human transportation in 1769, the vast and steady technological innovations over centuries have finally made vehicles evolved as an advanced, integrated and intelligent system on the road. Consequently, vehicle passengers have acquired persistently enhanced safety, comfort and convenience during the travel than ever before.

The newly emerged vehicular networking technology ushers in a new era of automobiles in which vehicles are enabled with wireless communication capability, and can be connected as a network on the wheel. Based on the smart vehicle model described in [3], the enabling components needed for vehicular communications include:

- *Communication facility*, also called on-board unit (OBU), to conduct wireless communications among vehicles and RSUs. The OBU operates on the DSRC radio which is specifically designed for short-range fast-motion vehicular communications. The details of DSRC will be discussed later in this chapter.

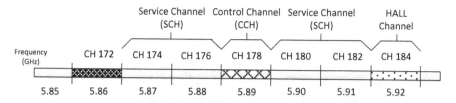

Fig. 1.1 DSRC channel spectrum

- *Positioning system*, such as a GPS receiver which can timely and accurately identify the location information, such as latitude, longitude and velocity of vehicles.
- *Sensors*, which include the autonomous ranging sensors such as the front and rear radars, cameras, and on-board sensors such as steering wheel angle sensors and wheel speed sensors. The vehicular radar, such as the microwave, infrared and ultrasonic radar, plays an important role for road safety by detecting and tracking the non-communication vehicles or obstacles. The on-board sensors are important to assist the efficient communications. For example, they can be used for motion and trajectory prediction and accordingly provide valuable information to vehicular network formation and optimization.
- *Computing, display and event data recorder facility* to enable on-board data processing, input/output and tracking to facilitate the vehicular communications.

Many of the above components, such as GPS, radars and on-board sensors, have already been mechanized and widely deployed in vehicular transportation.[1] In the near future, it can be envisioned that vehicles with the omni-capability of sensing, computing and communication will be pervasive. VANET will connect them as a ubiquitous network on the road.

1.1.1.2 Standardization

In 1999, the U.S. FCC allocated the 5.850–5.925 GHz frequency band for Dedicated Short-Range Communications (DSRC) dedicated to vehicular communications. The 75 MHz DSRC spectrum is divided into seven channels as shown in Figure 1.1. Among the seven DSRC channels, the fourth channel (CH 178), namely control channel, is exclusively used for the purpose of control, coordination and safety message broadcasting. The first channel (CH 172) is unused and the last (high availability low latency) channel (CH 184) is left for future use. The other four channels, namely service channels, are for IP-based infotainment applications.

---

[1] See "Technology on XTS, ATS Can Help Avoid Crashes", GM, 2012.3.27

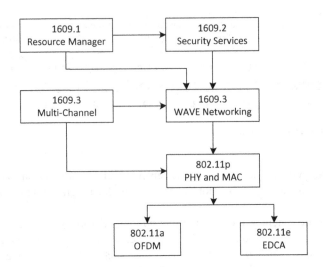

Fig. 1.2 WAVE protocol suite

The operation of DSRC-based VANET communications is standardized by the WAVE (wireless access for vehicular environment) protocol suite [4] as shown in Figure 1.2, where the IEEE 802.11p standard specifies the MAC and physical layers and the IEEE 1609 protocol suite specifies the upper-layer operations, as

- IEEE 1609.1 (WAVE Resource Manager) that describes the key components of WAVE system architecture and defines data flow and resource management protocols,
- IEEE 1609.2 (Security Services) that covers methods for securing WAVE management and application messages,
- IEEE 1609.3 (Networking Services) that defines network and transport layer services, including addressing and routing, etc.,
- IEEE 1609.4 (Multi-channel Operations) that defines the operations of the wireless transceiver over the multi-channel DSRC radio,
- IEEE 802.11p (MAC and Physical protocols) that specifies the MAC and physical layer operations of WAVE. The IEEE 802.11p protocol adopts the OFDM (Orthogonal Frequency Division Multiplexing) scheme in the physical layer which is similar to IEEE 802.11a. IEEE 802.11p targets a transmission range between 300 m and 1 km. It can provide the data transmission rates of 9, 12, 18, 24, and 27 Mbps to vehicles at the velocity of 0–60 km/h, and 3, 4.5, 6, 9, and 12 Mbps at the vehicle velocity of 60–120 km/h. In the MAC layer, the contention-based IEEE 802.11e EDCA (Enhanced Distributed Channel Access) scheme is adopted to support the QoS differentials of communications.

1.1.1.3 Industrial and Research Efforts on Vehicular Communications

As VANET is still in the infancy stage, a rich body of research has been developed to validate the effectiveness of wireless communications in the presence of high vehicular mobility.

Ott et al. [5] report the first measurement of the communication between a moving car with an external antenna and the roadside Wireless LAN access point (AP), namely drive-thru Internet. They show that using the off-the-shelf IEEE 802.11b device, a vehicle could maintain a connection to a roadside AP for around 600 m and transfer 9 MB of data at the velocity of 80 km/h using either TCP or UDP. Moreover, it is observed that the data transmission rate in each drive-thru presents a bell-shape curve with the very low rates when the vehicle enters and exits the AP coverage and transient high rate when the vehicle is close to the AP. Accordingly, the communication session within each drive-thru can be divided into three phases: entry phase, production phase, and exit phase, each of around 200 m in their experiments.

Bychkovsky et al. [6] have extended the experiments of drive-thru Internet in the metropolitan area. In the experiment, they make nine vehicles driven under the normal traffic conditions in the Boston metropolitan area, where each vehicle is equipped with a wireless enabled computer and communicates with the "open" residential WiFi AP during the driving. Through the analysis of measurement data collected in over 290 "drive hours", they report that the vehicles can successfully establish the AP connection and transmit data when vehicles move at the velocity ranging from 0 to 60 km/h, with the mean duration of connections to be 13 s and the mean duration between connections to be 75 s. Therefore, this experiment verifies the feasibility of communications between the moving vehicle to roadside AP regardless the velocity in the urban. Moreover, they show that using the current IEEE 802.11 protocol, it takes several seconds to setup the connection, which is intolerably long and significantly reduces the effective communication time of the connection. As a remedy to that, they propose an IP address caching scheme which can effectively reduces the AP association delay by by-passing the DHCP.

The prominent automobile corporations in the worldwide have also lunched important projects to promote vehicular broadband communications. For instance, Mercedes-Benz proposes to deploy the "InfoFuel" stations along the roads to fuel on-road vehicles with the high throughput Internet access using the IEEE 802.11a radio [7]. General Motors launched the "OnStar" to provide the subscription-based communications and in-vehicle security on the move [8], and Toyota builds "Toyota Friend" to facilitate the social communications among owners of Toyota cars [9].

1.2   Content Distribution in Vehicular Networks

The term "content" refers to general digital files, such as text messages, audio/video files, etc., which can be broadly classified as *small-* and *medium-size contents* of several or tens of megabytes, such as music files, picture images and text messages, and *large-volume or bulk contents*, such as video files of hundreds megabytes or several gigabytes.

1.2.1   Media Playback Pattern

Based on its playback pattern at the receivers, the content distribution applications can be generally categorized into two groups as:

- *Play-after-download*, in which the playback of the contents commences after the entire copy of the content file has been downloaded in the local cache.
- *Real-time streaming* (or *play-while-downloading*) where the content or media playback starts when the content is still downloading.

The play-after-download pattern can provide the best media playback quality to end users in terms of the visual quality and playback smoothness, as the intact content files have already been stored locally. Nevertheless, it typically incurs the long start-up delay to users who has to wait for the file download. The real-time streaming shortens the annoying start-up delay with immediate media presentation, but has the risk of frequent freezing of playback and severe video distortions due to the variable network delays and unpredictable packet losses.

The selection of the media playback pattern should not only cater to the specific content file size, but also conform to the available network throughput. For example, the playout of large volume contents should adopt the real-time streaming pattern in most scenarios to avoid the overly long start-up delays. Nevertheless, the medium-size contents, such as MP3 music files, should typically apply the download-then-play pattern for the guaranteed playback quality at the cost of modest start-up delay. The data contents like image and zip files are also typically required to be entirely downloaded before the presentation. However, it is important to note that the vehicular networks are highly dynamic with intermittent connectivity and intensively changing throughput. For example, as vehicles moving on the road with varying road traffics and densities of RSUs, the download throughput may change dynamically. As such, the content playback and usage pattern should adapt to the specific physical environments and download performance of vehicles along their trajectories. For example, when subscribing to watch a small video clip on Youtube but in a dynamic vehicular environment with the dramatically changing download rate, the play-after-download pattern may be more preferable than the real-time streaming pattern to avoid the possible interruptions and bad playback quality during the media presentation. When subscribing to access a social blog with a bundle of mixed files including audio, images and texts, in the vehicular networks with limited throughput, a modified play-after-download pattern may be preferable by selectively downloading a portion of the files in the bundle first and starting the presentation so as to strike the trade-off between the start-up delay and playback quality.

1.2.2 Design Criteria

It is also important to note that unlike the traditional research which largely focuses on the packet-level performance, such as the packet losses and packet delays during the communication, the emphasis of content distribution network design is to guarantee the session-level performance and optimize the user's perceived performance subject to the application specific QoS or Quality-of-Experience (QoE) requirements. For example, in the real-time video services, the design goal is to achieve the smooth video playback with modest start-up delays and high visual quality within the entire session of video playout. In the case of small- or media-size content distributions, the goal is to guarantee the integrity of the delivered contents and at the meantime reduce the transmission delay. The end-to-end transmission delay of each packet in the content is however not that critical.

In the context of vehicular networking, the content distribution network design should be tailored and adaptive to the specific road traffic and deployment environments. Specifically, different deployment scenarios, such as city urban, rural and highways, have entirely different features in the road traffic, vehicle mobility pattern and communication environment. The approaches of content distribution in different scenarios should therefore explore the specific features in different deployment environments. For example, in the urban area, there typical exist far more infrastructure devices, such as cellular base stations, WiFi access points, than other districts, like the rural and highways. As such, the key is to explore the opportunistic high-rate V2R connections to achieve the QoS guaranteed and cost-effective content distribution services to vehicles. In the highway scenario, although the infrastructure is typically sparsely deployed in order to save the deployment cost, as vehicles are moving in the one-dimensional linear topology with the relatively static mobility pattern, it is more likely to form the long-lasting V2V connections among vehicles and accordingly enable cooperations among vehicles to collaboratively distribute the contents. However, with the diverse mobility behaviors (such as velocities and acceleration/decelations) in the highway, how to identify the long-lasting V2V connections and utilize the diversity of vehicle mobilities are the key to form effective cooperations for content distributions on the highway.

1.2.3   Design Challenges and Goals

Content distribution and transmission typically has stringent QoS and QoE requirements. The efficient vehicular content distribution is challenged by the instinct
nature of vehicular networks from the following two aspects.

1.2.3.1   High Node Mobility

In VANETs, each communication node is a fast-moving vehicle. As a result, the
V2V and V2R communications are highly violative and susceptible to frequent
interruptions with the transient contact durations among nodes. For example, [6]
investigates on the download bandwidth of vehicles from the unplanned open
residential WiFi access points in Boston. It is shown that the plethora IEEE 802.11b
access points deployed in cities could provide vehicle nodes with the intermittent
and short-lived connectivity, yet high throughput when the connectivity is available.
The transient and intermittent download connectivity of vehicles inevitably renders
significant impairments to the on-top content distribution applications, such as live
and on-demand video streaming to vehicles.

Moreover, the dramatic changing connectivity and locations of vehicles lead to
the dynamic network topology. This dictates any content distribution protocols to
be resilient to the topology change and can self-heal quickly. Note that the mobility
of vehicles is pertained to the road layout and presents specific patterns in different
road environments. For example, the vehicle trajectory on the highway is basically
one-dimensional, and it is two-dimensional in the urban areas with dense street
intersections. The heterogeneity and diversity of the vehicle mobility and topology
patterns in VANETs thus indicate that there is no "one-for-all" solution to form
the content distribution networks. Instead, to address the network dynamics and
heterogeneities, the key is to *exploit the specific mobility patterns of vehicles in
different deploying environments, and accordingly adapt the content distribution
system and strategies to the particular topology patterns.*

1.2.3.2   Large Network Scale

The VANETs are typically of large-scale involving several hundreds or even
thousands of nodes communicating at the same time, such as all vehicles on
one highway section, in the downtown area of a city, or even in a region of
multiple cities. Therefore, *the content distribution networks must be scalable to the
network size.*

Moreover, due to the fast mobility of vehicles, the node density of the network
presents apparent spatio-temporal variations. As reported in [10], by analyzing
the real-world traces collected in Berkeley, California and Toronto, Canada, it is
shown that highway traffic drastically changes over time (of different scales) across

geographic locations. Also, it is intuitive that within a city, the vehicle density in the downtown area is typically greater than that in the outskirts. Even on the same road segment in a city, the vehicle density could fluctuate dramatically over time and is affected by the nearby traffic conditions. Therefore, *a practical design of the content distribution system needs to self-organize and be adaptive to the spatial- and time-varying road traffic.*

1.3 Research on Vehicular Content Distribution

On addressing the challenges of high node mobility and attaining the design goals to build efficient and scalable content distribution network, the previous literature mainly falls into two categories, namely infrastructure-based system and infrastructure-less system.

In the former category, the goal is to build an efficient and large-scale infrastructure which are dedicated to vehicular communication content distribution. The research proposal is to develop efficient algorithm with low time complexity to optimally manage the infrastructure equipment to provision guaranteed download performance to users. The latter relies on the vehicles to cooperatively distribute contents to each other. The research proposal is to develop the fully distributed algorithms which can fully explore the mobility pattern of individual vehicles and attain the global optimal download performance of the system in a fully distributed manner. In what follows, we will briefly introduce the related works within the two categories.

1.3.1 Infrastructure-Based Content Distribution

Frenkiel et al. [11] propose the Infostation in 1996 which can be used with the same purpose as RSUs to provide low-cost short-range communications to vehicles driving through. An Infostation is a wireless information kiosk deployed at roadside, building entrance, traffic lights and waiting areas at airports and train stations, etc. Users can download voice messages, faxes or e-mails when connecting to Infostations nearby. Within the Infostation framework, Yuen et al. [12] investigate on the non-cooperative content sharing in a mobile Infostation system. In [12], vehicles can download a file from the Infostation and swap among each other upon opportunistic contacts. By assuming that vehicles are selfish and reluctant to help the download of others, a tit-for-tat policy is applied in which the file is exchanged only when nodes upon the opportunistic contact have mutually interested contents.

Zhang et al. [13] focus on the design and optimization of RSU traffic control. Zhang et al. [13] considers the case when multiple vehicles connect to one RSU and contend the channel for retrieving their own contents from the RSU. In this scenario, a content scheduling algorithm at RSUs is develop which provides different service priorities based on the data size and service deadline of the contents. To be more specific, given two requests with the same deadline, the vehicle which requests for a smaller amount of data will be served first. Given two requests asking for data with same size, the vehicle with an earlier deadline will be served first. The algorithm is shown to outperform other scheduling algorithms, such as first-come-first-serve, first-deadline-first, and smallest data size first algorithms, to fully utilize the transient connection time of V2R communications.

Nandan et al. [14] describes AdTorrent which is content distribution application to disperse the advertising information pertaining to a local area. In AdTorrent, static wireless digital billboards are deployed on the roadside, continually pushing the advertising contents, e.g., hotel virtual tours, movie trailers, etc., to the driving through vehicles. Among the vehicles, the advertising contents are then swapped in a peer-to-peer style through fully distributed V2V communications.

Huang et al. [15] propose to deploy buffer storages in the city to enable the content retrieval of vehicles. Assuming that an infrastructure grid has been deployed in the city and without considering any V2V communications, [15] develops an optimal content replication scheme which determines the optimal locations of content files in the infrastructure grid based on the trip path of vehicles. As such, along the trip of vehicles, [15] strives to maximizing the delivery ratio of contents to vehicles using the infrastructure.

1.3.2 Infrastructure-Less Content Distribution

Nandan et al. [16] propose SPAWN (swarming protocol for vehicular ad-hoc networks), an infrastructure-less content distribution protocol based on V2V communications. SPAWN enables vehicles to cooperatively communicate with each other for content retrieval and sharing. In SPAWN, a file is chopped into multiple pieces and swapped among vehicles in a BitTorrent (BT) style. Using the proximity-driven piece selection, SPAWN exploits the location information of vehicles for piece selection and download which can outperform the traditional rarest first scheme used in BT.

Targeting to the same application scenario, Li et al. propose CodeOn [17], which applies the network coding to facilitate the multicast content distribution on the highway. In CodeOn, the interest contents are first encoded using the symbol level network coding, which not only enhances the diversity of content blocks, but also makes the system robust to the severe packet loss due to the harsh wireless channels. Within this framework, according to the amount of useful content blocks each vehicle has, an optimal relay selection protocol is presented to maximize the downloading rate of the system. Ye et al. [18], Yang et al. [19], Yan et al. [20] also investigate on the network coding based vehicular content distribution network.

Zhang et al. propose Roadcast [21] in which two functions are provided to facilitate the efficient V2V content distribution. Firstly, a vector space model (VSM) based content search is enabled to help vehicles hunt for the interested files from the ocean of content files stored in the distributed vehicles. Secondly, a popularity-aware content distribution protocol is proposed to minimize the integrated download delay of files. Specifically, assuming the fixed amount storage in the network, [22] shows that the minimum download delay of files is achieved when the number of file copies in the network is proportional to square-root of their popularity in the system steady state. In the light of [22], Roadcast assigns each file a cost value which is evaluated based on the popularity, size and estimated download delay. According to its cost value, files are evicted from the node buffer with different priorities, and finally the network is shown to converge to a stable state following the "square-root" law as in [22].

Zhang et al. propose V-PADA [23] a vehicle-platoon-aware scheme to utilize the clustering/platoon effect of vehicles on the highway for content distribution. In specific, it is shown in [24] that vehicles on the highway tend to form clusters on the road. As such, vehicles within the same cluster can cooperative store contents and collaboratively distribute contents. Inspired by this idea, V-PADA deploys the stochastic time series analysis to predict the movement of vehicles in the same platoon and based on the mobility of vehicles to determine the content replications. As such, contents are optimally replicated at distributed vehicles in a stable platoon to minimize the download costs of vehicles in the platoon.

1.4 Organization of the Monograph

The rest of this monograph is organized as follows:

Chapters 2 and 3 focus on the discussion of distributing small- and medium-size content to vehicles using infrastructure-based and infrastructure-less systems, respectively.

In Chap. 2, we describe to build a city-wide infrastructure system which deploys a multitude of wireless buffer devices on the roadside and provide opportunistic download to vehicles when driving through the coverage of the devices. To ensure that vehicles can download with guaranteed QoS, we describe a distributed algorithm which allows the individual devices to determine the replication of contents in their buffer. The presented infrastructure explores the distributed V2V communications and can be incorporated with the other access methods, such as cellular networks and city-wide WiFi networks as a complement for low-cost and high-rate content distribution.

In Chap. 3, we consider the content transmissions over the highly dynamic V2V communications. Note that content files need to be transmitted in their entirety to guarantee the successful presentation by the upper-layer applications. The V2V communications, however, are susceptible for frequent interruptions due to the diverse vehicle mobility. Therefore, content transmissions over the dynamic V2V communications may suffer from the fragment content issues in which the content files could only be partially transmitted during the short-lived connection time between sender and receiver. The fragment contents could be very harmful to waste the precious bandwidth of vehicles. On addressing this issue, we describe the integrity-oriented content transmission framework. The proposal can evaluate the data volume transmitted over the dynamic V2V connections, and filter the suspicious transmission requests which are unlikely to be finished within the connection time and accordingly avoid fragment contents.

Chapter 4 considers the case when a large-volume content is distributed to in-motion vehicles and the content is played at the same time of downloading, like on-demand video streaming. In this case, with vehicles opportunistically connecting to roadside communications infrastructure and downloading with dramatically changing throughput and delays, Chap. 4 describes the adaptive content playout scheme which combats with network dynamics in order to provide the smooth content presentation at the upper-layer application.

Chapter 5 concludes this monograph.

References

1. T. . Globe and M. Update, "Canadians' Internet Use Exceeds TV Time," Mar. 2010.
2. P. Bouvard, L. Rosin, J. Snyde, and J. Noel, "The Arbition National In-Car Study," *Arbitron/Edison Media Research*, Dec. 2003.
3. J. P. Hubaux, S. Capkun, and J. Luo, "The Security and Privacy of Smart Vehicles," *IEEE Security & Privacy*, vol. 2, pp. 49–55, May 2004.
4. Y. L. Morgan, "Notes on DSRC & WAVE Standards Suite: Its Architecture, Design, and Characteristics," *IEEE Communications Surveys & Tutorials*, vol. 12, pp. 504–518, Fourth Quarter 2010.
5. J. Ott and D. Kutscher, "Drive-Thru Internet: IEEE 802.11b for "Automobile" Users," in *Proc. of IEEE Infocom*, 2004.
6. V. Bychkovsky, B. Hull, A. Miu, H. Balakrishnan, and S. Madden, "A Measurement Study of Vehicular Internet Access Using In Situ Wi-Fi Networks," in *Proc. of ACM MobiCom*, 2006.
7. J. Angel, "Mercedes-Benz Demos Wireless Network," 2001.
8. G. Motors, "On Star," accessed in Jan. 2013.
9. G. Wong, "Toyota to Launch 'Toyota Friend': a Car-Owner Social Network," May 2011.
10. F. Bai and B. Krishnamachari, "Spatio-Temporal Variations of Vehicle Traffic in VANETs: Facts and Implications," in *Proc. of ACM VANETs*, 2009.
11. R. H. Frenkiel and T. Imielinski, "Infostations: The Joy of Many-Time Many-Where Communications," *Journal on Mobile Computing*, 1996.
12. W. H. Yuen, R. D. Yates, and S. C. Mau, "Exploiting Data Diversity and Multiuser Diversity in Noncooperative Mobile Infostation Networks," in *Proc. of IEEE Infocom*, 2003.
13. Y. Zhang, J. Zhao, and G. Cao, "On Scheduling Vehicle-Roadside Data Access," in *Proc. of ACM VANET*, 2007.
14. A. Nandan, S. Das, B. Zhou, G. Pau, and M. Gerla, "AdTorrent: Digital Billboards for Vehicular Networks," in *Proc. of IEEE/ACM V2VCOM*, 2005.
15. Y. Huang, Y. Gao, K. Nahrstedt, and W. He, "Optimizing File Retrieval in Delay-Tolerant Content Distribution Community," in *Proc. of IEEE ICDCS*, 2009.
16. A. Nandan, S. Das, G. Pau, M. Gerla, and M. Y. Sanadidi, "Co-Operative Downloading in Vehicular Ad-Hoc Wireless Networks," in *Proc. of IEEE/IFIP WONS*, 2005.
17. M. Li, Z. Yang, and W. Lou, "Codeon: Cooperative Popular Content Distribution for Vehicular Networks using Symbol Level Network Coding," *IEEE Journal on Selected Areas in Communications*, vol. 29, pp. 223–235, Jan. 2011.

18. F. Ye, S. Roy, and H. Wang, "Efficient Data Dissemination in Vehicular Ad Hoc Networks," *IEEE Journal on Selected Areas in Communications*, vol. 30, pp. 769–779, May 2012.

19. Z. Yang, M. Li, and W. Lou, "Codeplay: Live Multimedia Streaming in VANETs Using Symbol-Level Network Coding," in *Proc. of IEEE ICNP*, pp. 223–232, 2010.

20. Q. Yan, M. Li, Z. Yang, W. Lou, and H. Zhai, "Throughput Analysis of Cooperative Mobile Content Distribution in Vehicular Network using Symbol Level Network Coding," *IEEE Journal on Selected Areas in Communications*, vol. 30, pp. 484–492, Feb. 2012.

21. Y. Zhang, J. Zhao, and G. Cao, "Roadcast: a Popularity Aware Content Sharing Scheme in VANETs," in *Proc. of IEEE ICDCS*, 2009.

22. E. Cohen and S. Shenker, "Replication Strategies in Unstructured Peer-to-Peer Networks," *ACM SIGCOMM Computer Communication Review*, vol. 32, pp. 177–190, Oct. 2002.

23. Y. Zhang and G. Cao, "V-PADA: Vehicle-Platoon Aware Data Access in VANETs," *IEEE Transactions on Vehicular Technology*, vol. 60, pp. 2326–2339, June 2011.

24. N. Wisitpongphan, F. Bai, P. Mudalige, V. Sadekar, and O. Tonguz, "Routing in Sparse Vehicular Ad Hoc Wireless Networks," *IEEE Journal on Selected Areas in Communications*, vol. 25, pp. 1538–1556, Oct. 2007.

Chapter 2
Medium-Size Content Distribution Using an Infrastructure-Based Approach

Our lives in the city are exposed to a rich mixture of different types of information and contents everyday, such as store flyers, newspaper, advertisements and social information. In the context of vehicular applications, the road-related contents such as traffic alert and road condition reports are also important to be timely distributed to drivers. In this chapter, we consider the distribution of medium-size contents through an infrastructure-based approach in urban vehicular networks.

2.1 Introduction

While being actively pursued for years, the real-world large-scale deployment of city-wide broadband vehicular communications in general and vehicular content distribution services in specific is still not practical and fraught with many fundamental challenges.

This mainly attributes to the lack of an efficient accessing approach on providing ubiquitous, high-rate yet low-cost connections to vehicles. The development of RSU is still in the research stage; the large-scale deployment of RSUs tends to be slow process as it is not only monetarily expensive but also may conflict and need to compete with existing access networks in cities. Using the traditional 3G/4G cellular networks, not only the aggregate bandwidth per user is very limited as a large number of users need to share wireless resource concurrently, but also the usage cost per user is high. An alternative approach is by exploring city-wide WiFi hotspots for high-rate services at the low price. However, sparsely distributed in the city [1, 2] with limited coverage individually and stringent access control, WiFi hotspots can hardly provide ubiquitous connectivity to vehicles. Moreover, originally designed for static indoor applications, WiFi hotspots are not optimized for highly mobile vehicular communications [3]. Another plausible solution is by exploring inter-vehicle communications. While collaborative inter-vehicular communications can boost the system capacity, purely relying on the vehicle-to-vehicle (V2V)

T.H. Luan et al., *Enabling Content Distribution in Vehicular Ad Hoc Networks*,
SpringerBriefs in Computer Science, DOI 10.1007/978-1-4939-0691-8_2,
© The Author(s) 2014

communications is insufficient to provide the reliable and high-rate data services to users due to harsh channel conditions and unreliable inter-vehicle connections [4,5]. As reported in [6], the throughput of inter-vehicular communications is observed to be at most one fifth of the throughput of vehicle-to-infrastructure communications. To summarize, in order to bring vehicular communications and vehicular content distributions from lab concept to commercial reality, a novel, practical and scalable infrastructure which offsets the weaknesses of traditional accessing approaches and dedicated to vehicular communications with reserved communication resources is necessary.

In this chapter, we describe a proposal presented in [7] towards this goal. This is by building a new city-wide cost-effective infrastructure which can well coexist with existing access networks and provide dedicated communication facilities to enable the QoS guaranteed content and information distribution services to vehicles.

Note that to build a large-scale infrastructure in city is typically a daunting task, if not impossible, due to the high deployment and maintenance cost. The foremost issue is therefore to answer the question: how to construct a *practical* and *cost-effective* infrastructure which not only bypasses physical installations and investment obstacles, but also incurs the minimum monetary wireless bandwidth expense for individual users. Assuming that such an infrastructure is formed, another question to answer is how to manage the individual equipment in the infrastructure so as to provide the guaranteed download performance to vehicles.

We address the above two questions in sequence in the remainder of this chapter. In specific, we first provide an overview description of the infrastructure, including its basic component, operation and formation process, and answer the first question. After that, we answer the second question by describing the detailed model and algorithm design of the infrastructure, including the mathematical model of network, algorithm design for infrastructure management and the content distribution prototype. Lastly, we present simulation results to show the performance of the described infrastructure, and close this chapter with a summary and discuss on the future works.

2.2 Overview of the Described Infrastructure

2.2.1 Basic Component

The described infrastructure is composed of a multitude of wireless buffer devices deployed on the roadside, namely roadside buffers (RSBs). Each RSB is equipped with a wireless transceiver operating on the dedicated short-range communication (DSRC) radio, and can communicate with nearby vehicles using the vehicle-to-infrastructure communications. The RSB can selectively retrieves content files from the vehicle drive-through its coverage and disseminate the cached files to vehicles upon their requests. To summarize, RSBs are different from traditional RSUs from the following two aspects:

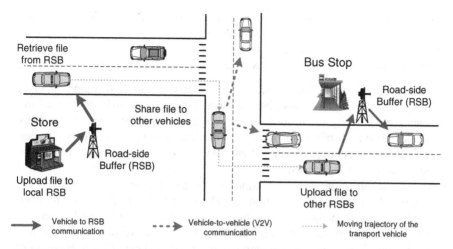

Fig. 2.1 Content distribution through the described infrastructure

- *Internet Connectivity*: RSBs are not necessarily connected to the Internet, whereas RSUs are typically connected to the Internet so as to provide the Internet access to vehicles.
- *Buffer Storage*: RSBs have a local buffer to store contents. They are able to retrieve contents from vehicles in the communication range and in the meantime upload the coached contents to vehicles per their requests. The traditional RSUs are similar to WiFi APs which provide the straight Information connections to vehicles and are not equipped with the large-volume buffer to cache contents and serve the distribution.

Figure 2.1 shows a motivating example in which a grocery store intends to distribute its recent flyers to customers in the city. To do so, the flyers are first uploaded to one or multiple RSBs near the store. The RSBs are then responsible for distributing the content files (flyers) on the fly to vehicles driving through the area and let the vehicles spread the flyers to other RSBs and vehicles across the city.

2.2.2   Formation Process of the Infrastructure

Unlike the conventional centralized system (e.g., cellular base stations), the described infrastructure (i.e., RSBs) is distributedly deployed, owned and managed by separate entities. For example, a grocery store or shopping mall may deploy the RSB in its parking lot to periodically broadcast the flyers as in Figure 2.1. A movie theater may install the RSB to distribute the latest movie tailors to the public. The distributed RSBs deployed by separate entities collectively form the infrastructure network. In other words, the formation of the RSB infrastructure relies on the contributions of separate individuals in the city with the shared investment and maintenance work of the devices.

The distributed deployments of RSBs have the following features:

- *Cheap and Easy to Install*: the RSBs are cheap and light-weight devices composed of a wireless transceiver and small buffer. They can be configured and managed wirelessly, requiring no complex and expensive cabling work. As RSBs are deployed to distribute the local contents generated by their owners, they are not necessarily connected to the Internet. As such, once deployed, RSBs incur no bandwidth cost to their owners.
- *Easy to Manage*: the content distribution and buffer management of RSBs are purely self-organized which autonomously adapt to the time-varying network conditions (e.g., the density of vehicle traffic, buffer availability) [8] and are tailored to meet the file download demands. Therefore, except to using wireless connections, the owners are not required to get involved in any further operation.
- *Profitable*: The RSBs can bring commercial benefits to their owners by distributing the advertisements or other information to the public.

To summarize, the RSBs distributedly deployed in the city can provide dedicated storage and communication capacity to enable content distributions to vehicles. Moreover, by relying on the fast vehicles to transport contents among RSBs, and making RSBs selectively retrieve contents from vehicles to cache and redistribute, the entire infrastructure is designed to achieve a global optimal goal in a fully distributed manner.

2.3   Network Design and Optimization

Suppose that a large-scale RSB infrastructure can be formed in a fully distributed manner as described in the previous section, in this section, we describe the detailed algorithm and protocol design of the RSB infrastructure so as to enable the efficient and QoS guaranteed content distribution to vehicles. To attain this goal, we start by presenting the system model, including the RSB modelling, mobility of vehicles and the utility function of vehicular users. Based on the developed system model, the network design is then formulated as an optimization problem. The main notations used are summarized in Table 2.1.

Table 2.1 Summary of notations

Notations	Description
$\mathbf{R}$	Set of RSBs in the region of interest
$\mathbf{F}$	Set of files published for download in the region of interest
$\mathbf{P}$	Popularity profile of files, where each element p_i, $i \in \mathbf{F}$, represents the probability that a user subscribes to download file i upon each download request
$\mathbf{A}$	Availability profile of files, where each element a_i, $i \in \mathbf{F}$, represents the portion of users which have file i stored in the local buffer
$\mathbf{B}$	Caching profile of files, where each element b_i, $i \in \mathbf{F}$, represents the probability that file i is stored in a RSB
C_{V2R}	Communication data rate between RSBs and vehicles
C_{V2V}	Communication data rate of V2V communications
B_{R}	Buffer size of a RSB
$U(\cdot)$	Utility function of vehicular users
$\mathscr{U}$	Global network utility (overall performance of network to optimize)
r	Download throughput of vehicles when vehicles are outside the communication of RSBs
R	Download throughput of vehicles when vehicles are inside the communication of RSBs
n	Number of vehicles which are able to transmit to the tagged node, or contend the channel for transmission with the tagged node
κ_i	Number of blocks in file i
τ_i	Mean download delay of file i
$1/\lambda$	Mean sojourn time of vehicles inside the communication range of RSBs
$1/\mu$	Mean sojourn time of vehicles outside the communication range of RSBs
$1/\delta$	Mean file block download time of vehicles outside the communication range of RSBs
$1/\gamma$	Mean file block download time of vehicles inside the communication range of RSBs
$\Gamma(m,k)$	Mean first passage time of Markov process from state (m,k) to state $(\cdot, \kappa_i)$
$|\cdot|$	Cardinality of set
$\langle\cdot\rangle$	Mean value of a random variable
$\mathrm{Var}(\cdot)$	Variance of a random variable

2.3.1 Network Model

2.3.1.1 Model of RSBs

We consider the city as a bounded region where a set **R** of RSBs are randomly deployed. Note that with different building environments and diverse communication capabilities, RSBs at different locations would have different radio coverage. Within their communication coverage, we consider RSBs to have the same data transmission rate to vehicle nodes, denoted by C_{V2R}. In this chapter, we allow vehicles to communicate with each other to cooperatively disseminate the downloaded contents to each other. Let C_{V2V} denote the data transmission rate of V2V communications. Each vehicle is equipped with a single-radio transceiver and communicate to only one node at each time, which is same as [9]. We make $C_{V2R} > C_{V2V}$, and vehicles prefer to downloading from RSBs if RSB connections are available. This is a working assumption as RSBs tend to have higher transmission rate than V2V communications due to the ample power energy and better channel conditions when mounted high [6].

2.3.1.2 Model of Vehicle Mobility

The mobility of each vehicle node is represented by an *on–off* process based on its connectivity to RSBs: a vehicle node is in state 0 if it is outside the coverage of any RSB; otherwise, it is in state 1. Due to the random radio coverage and deployment locations of RSBs, we model the sojourn time of vehicles in state 1 and state 0 by the unpredictable, memoryless and continuous-time setting, following an exponential distribution with the mean value $1/\lambda$ and $1/\mu$, respectively.

2.3.1.3 Model of Files

Let **F** denote the integrated set of content files available for download in the region of interest. Throughout the work, each RSB is assumed to be manipulated by a distinct owner; the owner uploads contents to its RSB at periodic intervals following the exponential distribution with the mean Δ. RSBs have homogenous buffer size[1] which is denoted by L. When the buffer of RSBs overflows with excessive file uploading from vehicles, the oldest files stored in the RSB will be evicted.[2]

Throughout the work, we focus on the design of RSBs and assume that the buffer management at vehicle nodes are predefined and out of the control. In specific, the

---

[1] In practice, the RSBs would be produced by the same vendor with equal buffer size.

[2] It is interesting to investigate on the impacts of different buffer management schemes, e.g., least frequently used (LFU) and least recently used (LRU) on the download performance, which however is out the scope of this work.

vehicles could have heterogenous and limited sized buffer storage, and randomly select files to evict if their buffer overflows. The pattern of V2V content swap is also predefined which could follow existing schemes, such as SPAWN [10].

With new files being continually published at distributed RSBs and old files being evicted from the network, **F** dynamically changes over time. In the network, each file is characterized by a three-tuple, including file blocks, popularity and availability, defined as follows.

File Blocks: each content file in the network is divided into multiple non-overlapping file blocks for delivery. In order to finish downloading a file, a vehicle node must collect all blocks of the requested file from either RSBs or other vehicles with the file stored. A vehicle node can only redistribute a file to the others after it has the entire file downloaded and recovered.[3] Let κ_i denote the number of blocks of file i, where $i \in \mathbf{F}$. For ease of analysis, we assume that all the blocks of files have equal size. With files having different numbers of blocks, they are heterogenous in size. For computation simplicity, L, C_{V2R} and C_{V2V} are normalized by the block size.

File Popularity and Availability: besides the number of file blocks, each content file in the network is characterized by another two parameters, namely popularity and availability.

Definition 2.1. The popularity p_i of file i, where $i \in \mathbf{F}$, represents the probability that a vehicle subscribes to download file i upon each download request which it issues. The *popularity profile* of **F** is a $1 \times |\mathbf{F}|$ probability vector $\mathbf{P} = \{p_i; i \in \mathbf{F}\}$, where $|\cdot|$ indicates the cardinality of set.

Definition 2.2. The availability a_i of file i, where $i \in \mathbf{F}$, represents the probability that a randomly selected vehicle has file i cached in its buffer. The *availability profile* of **F** is a $1 \times |\mathbf{F}|$ probability vector denoted by $\mathbf{A} = \{a_i; i \in \mathbf{F}\}$.

Note that since **F** is dynamically changing over time, the popularity profile **P** and availability profile **A** are also varying over time. In this case, RSBs stochastically select files in **F** to cache in their buffer following the caching profile defined below.

Definition 2.3. Let b_i denote the probability that a randomly selected RSB has file i stored in its buffer. The *caching profile* of **F** is a $1 \times |\mathbf{F}|$ probability vector denoted by $\mathbf{B} = \{b_i; i \in \mathbf{F}\}$.

2.3.1.4 Mean Download Delay of Files

The performance of the network is characterized by the mean download delay of files. Let τ_i denote the mean download delay of file i which starts when a download request of file i is issued by a vehicle until the the vehicle finishes downloading all the blocks of file i. Given the distribution of RSBs and density of vehicle nodes in

---

[3]It can be extended by allowing vehicles to redistribute file blocks as long as certain blocks are downloaded in entirety. We study the simplest case and leave the extension for future works.

the region of interest, the download delay τ_i is dependent on the availability of file i at RSBs, represented by b_i, and vehicles, represented by a_i.

2.3.2 Network Utility Function

For each file i, we assume that there is an underlying utility function $U_i(\tau_i)$ that specifies the satisfaction of vehicular users on the download of file i provided the download delay τ_i. Moreover, it is nature to assume that $U(\tau_i)$ is a monotonically decreasing function of τ_i, i.e., reducing the download delay τ_i would monotonically increase the user's utility of file i.

The RSB infrastructure is designed to maximize a global network utility function $\mathscr{U}$ which represents the integrated utilities of vehicles. In general cases, it can be expressed as a weighted sum of individual user utilities over all files, mathematically,

$$\mathscr{U} = \sum_{i \in \mathbf{F}} w_i U(\tau_i), \qquad (2.1)$$

where $w_i, i \in \mathbf{F}$, is a given positive weight. With different concerns, the network utility can be adapted to achieve different design goals, as following examples:

2.3.2.1   User-Centric Content Distribution

In this scenario, by tuning the weighting factor of each file equal to the corresponding file popularity, the RSB infrastructure targets to optimize the user's download experience by maximizing the average user satisfaction on the file dissemination. Mathematically, the network utility is given as

$$\mathscr{U} = \sum_{i \in \mathbf{F}} p_i U(\tau_i). \qquad (2.2)$$

2.3.2.2   Content-Centric Content Distribution

The weighting factor w_i can be set to a predefined value which reflects the importance of file i. For example, breaking news, important software update, etc., can be assigned with the large weighting factors and accordingly attain high priorities to be stored in RSBs. This ensures those important files to be vastly stored and ubiquitously available.

2.3.2.3   Cost-Centric Content Distribution

A practical concern of the RSB infrastructure is the hardware cost of RSBs. With larger buffer storage of RSBs, more files can be cached in each RSB, rendering reduced download delay to users; nevertheless, it increases the cost of RSB hardware accordingly. Motivated by this concern, the network utility can be modified by introducing the cost function in (2.1) to strike a trade-off between the network performance and investment cost, as

$$\mathscr{U} = \sum_{i \in \mathbf{F}} w_i U\left(\tau_i\right) - \mathbf{C}\left(L\right), \tag{2.3}$$

where $\mathbf{C}\left(L\right)$ represents the hardware cost of RSBs which is a non-decreasing function of the buffer size L. In practice, $\mathbf{C}\left(L\right)$ cannot be evaluated by $\mathbf{C}\left(\sum_{i \in \mathbf{F}} b_i \kappa_i\right)$ instead, where $\sum_{i \in \mathbf{F}} b_i \kappa_i$ represents the mean usage of RSB buffer storage. As such, the three designs of the network as aforementioned can be solved using the unified formulation as described below.

2.3.3   Network Formulation

Given the network model introduced above, each RSB distributedly determine the optimal caching profile $\mathbf{B}$ of files to attain the maximal network utility $\mathscr{U}$, mathematically,

$$\begin{aligned} maximize \quad & \mathscr{U} \\ subject\ to: \quad & \Pr(X \geq L) \leq \varepsilon, \\ & b_i \in [0,1], \quad i \in \mathbf{F}, \end{aligned} \tag{2.4}$$

where X denotes the usage of the RSB buffer storage at any time, and $0 < \varepsilon << 1$ is a predefined constant. The constraint of (2.4) specifies that the overflow probability of each RSB should be no larger than ε.

Lemma 2.1. *Let* X_i, $i \in \mathbf{F}$, *be an independent binary random variable with the probability mass function*

$$\Pr\left(X_i = 1\right) = b_i, \quad \Pr\left(X_i = 0\right) = 1 - b_i,$$

which denotes whether file i *is stored in a RSB or not. For* $X = \sum_{i \in \mathbf{F}} X_i \kappa_i$ *with* $\kappa_i > 0$, *which denotes the usage of RSB buffer storage, we have* $E\left(X\right) = \sum_{i \in \mathbf{F}} b_i \kappa_i$. *By denoting* $v = \sum_{i \in \mathbf{F}} b_i \kappa_i^2$, *we have*

$$\Pr\left(X \geq E\left(X\right) + \psi\right) \leq \exp\left(-\frac{\psi^2}{2\left(v + \kappa\psi/3\right)}\right) \tag{2.5}$$

where $\kappa = \max\{\kappa_i; i \in \mathbf{F}\}$.

Proof. Refer to [11] (p. 25).

Theorem 2.1. *Given the network modeling, the constraint of (2.4) is achieved when*

$$E(X) \leq L - \kappa \frac{2}{3} \log \varepsilon - \sqrt{\kappa^2 \frac{4}{9} \log^2 \varepsilon - 2\kappa L \log \varepsilon}. \qquad (2.6)$$

Proof. According to Lemma 2.1, given the caching profile **B**, we have that the usage of RSB buffer storage X satisfies (2.5). As $v = \sum_{i \in \mathbf{F}} x_i \kappa_i^2 \leq \kappa E(X)$, where v and κ are as defined in Lemma 2.1. By substituting $v \leq \kappa E(X)$ into (2.5), we have

$$P(X \geq E(X) + \psi) \leq \exp\left(-\frac{\psi^2}{2(v + \kappa\psi/3)}\right)$$

$$\leq \exp\left(-\frac{\psi^2}{2(\kappa E(X) + \kappa\psi/3)}\right).$$

By assuming $L = E(X) + \psi$ and substituting it into (2.5), we have

$$\Pr(X \geq L) \leq \exp\left(-\frac{(L - E(X))^2}{2(\kappa E(X) + \kappa(L - E(X))/3)}\right).$$

As such, the constraint of (2.4) can be achieved if

$$\exp\left(-\frac{(L - E(X))^2}{2(\kappa E(X) + \kappa(L - E(X))/3)}\right) \leq \varepsilon. \qquad (2.7)$$

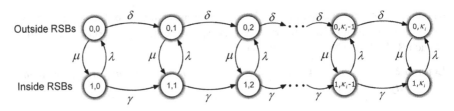

Fig. 2.2 State space and transitions of the two-dimensional Markov process

By solving (2.7), we have that the constraint of (2.4) is satisfied if

$$E(X) = \sum_{i \in \mathbf{F}} x_i \kappa_i \leq L - \kappa \frac{2}{3} \log \varepsilon - \sqrt{\kappa^2 \frac{4}{9} \log^2 \varepsilon - 2\kappa L \log \varepsilon}.$$

Denote by $\mathscr{L} = L - \kappa\frac{2}{3}\log\varepsilon - \sqrt{\kappa^2\frac{4}{9}\log^2\varepsilon - 2\kappa L\log\varepsilon}$. With Theorem 2.1, (2.4) can be modified as

$$
\begin{aligned}
\text{OPT} \quad & maximize \quad \quad \mathscr{U} \\
& subject\ to: \sum_{i\in\mathbf{F}} b_i\kappa_i \leq \mathscr{L}, \\
& \quad\quad\quad b_i \in [0,1], \quad i \in \mathbf{F}.
\end{aligned} \tag{2.8}
$$

In (2.8), with **A** and **P** provided, tuning the caching profile **B**, i.e., content replications in RSBs, will adapt the download delay τ_i of each file i and accordingly lead to different network utility $\mathscr{U}$. In this work, our goal is to determine the solution of (2.8) in a distributed manner.

2.3.4 Evaluation of File Download Delay

To solve (2.8), the foremost issue is to identify the quantitative relation between the file download delay and the caching profile **B**. To this end, we randomly select a vehicle node from the network (referred to as the tagged node) and evaluate its download delay of file i. Specifically, based on the system model described in Sect. 2.3.1, we represent the tagged vehicle node by a two-dimensional Markov process $(M_i(t), K_i(t))$. Here, $M_i(t) \in \{0,1\}$ represents the mobility of the vehicle node according to the on–off model described in Sect. 2.3.1, and $K_i(t) \in \{0,1,\ldots,\kappa_i\}$ represents the number of file blocks that the tagged node has downloaded until time t. Figure 2.2 shows the state space of the Markov process and all the non-null transitions. In what follows, we evaluate the transition rates of the Markov process according to the locations of the tagged node.

2.3.4.1 Tagged Vehicle Outside the Coverage of RSBs

When the tagged node is outside the coverage of any RSBs, it can only download from nearby vehicles using the V2V communications; at each time, we refer to the set of vehicles which are within the communication range of the tagged node as the neighbor nodes. Let n denote the number of neighbor nodes of the tagged node; n is a random variable, and let $\langle n \rangle$ and $\mathrm{Var}(n)$ denote its mean and variance, respectively. Assuming that the vehicular network has an ideal MAC where the channel airtime is fairly shared among the nearby vehicles, the throughput of the tagged node using the V2V communication is a function of n as,

$$
r = \frac{C_{\text{V2V}}}{n+1} Q_i(n), \tag{2.9}
$$

where $Q_i(n) = 1 - (1-a_i)^n$, representing the probability that at least one neighbor node of the tagged node has file i stored in its buffer, or equivalently, the probability that the tagged node can retrieve file i from its neighbor nodes. $(n+1)$ in (2.9) represents the number of vehicles fairly sharing the channel.

Let δ denote the transition rate from state $(0, K_i(t))$ to state $(0, K_i(t) + 1)$, where $K_i(t) \in \{0, 1, \ldots, \kappa_i - 1\}$, as shown in Figure 2.2. Assuming that the download time of one block using the V2V communication follows the exponential distribution, δ is equal to the mean V2V communication throughput $\langle r \rangle$ where r is specified in (2.9). Taking the expectation on n in both sides of (2.9), we approximate $\langle r \rangle$ using the second order Taylor series approximation as shown in Lemma 2.2.

Lemma 2.2. *With the second order Taylor approximation, we have*

$$\langle r \rangle \approx r|_{\langle n \rangle} + \frac{1}{2} \mathrm{Var}(n) \left. \frac{d^2 r}{dn^2} \right|_{\langle n \rangle}. \tag{2.10}$$

Proof. By applying the Taylor series expansion, the second order approximation of r as a function n can be represented as

$$r \approx r|_{\langle n \rangle} + (n - \langle n \rangle) \left. \frac{dr}{dn} \right|_{\langle n \rangle} + \frac{1}{2} (n - \langle n \rangle)^2 \left. \frac{dr^2}{dn^2} \right|_{\langle n \rangle}. \tag{2.11}$$

By taking the expectation on both sides of (2.11) with respect to n, we have

$$\langle r \rangle \approx r|_{\langle n \rangle} + \frac{1}{2} \mathrm{Var}(n) \left. \frac{dr^2}{dn^2} \right|_{\langle n \rangle},$$

where $\mathrm{Var}(n)$ denotes the variance of n.

2.3.4.2 Tagged Vehicle Inside the Coverage of RSBs

In this case, the tagged node can download the demanded blocks from either neighbor vehicles or the RSB. We assume that the tagged vehicle would select to download from RSBs with high priority, if the connected RSBs have the desired file stored; otherwise, it would download from neighbor vehicle nodes. This is because that RSBs have the greater communication capacity than vehicles [6]. In this scenario, given that file i is stored at the RSB with probability b_i, the download throughput of the tagged vehicle in this scenario is

$$R = b_i \frac{C_{\mathrm{V2R}}}{n + 1} + (1 - b_i) r. \tag{2.12}$$

The first component on the right-hand-side of (2.12) represents the download rate from the RSB with the ideal MAC applied, and the second component on the right-hand-side of (2.12) represents the download rate using the V2V communications given that with probability $(1 - b_i)$ that the RSB does not have the desired file i stored.

Let γ denote the transition rate from the state $(1, K_i(t))$ to the state $(1, K_i(t) + 1)$, where $K_i(t) \in \{0, 1, \ldots, \kappa_i - 1\}$, as shown in Figure 2.2. Similar to the previous case, we assume that the download time of one block inside the RSB follows the exponential distribution. Therefore, we have γ equal to $\langle R \rangle$ with R shown in (2.12). $\langle R \rangle$ can be approximated with the second order Taylor approximation as in Lemma 2.3.

Lemma 2.3. *With the second order Taylor approximation, we have*

$$\langle R \rangle \approx b_i \Phi + (1 - b_2) \langle r \rangle , \tag{2.13}$$

where $\Phi = C_{V2R} \left(\frac{1}{\langle n \rangle + 1} + \frac{\mathrm{Var}(n)}{(\langle n \rangle + 1)^3} \right).$

Proof. Similar to the proof of Lemma 2.2, by applying the Taylor series expansion, the second order approximation of R as a function of n represented as

$$R \approx b_i \left(G(\langle n \rangle) + (n - \langle n \rangle) \left. \frac{dG(n)}{dn} \right|_{\langle n \rangle} \right. \tag{2.14}$$
$$\left. + \frac{1}{2} (n - \langle n \rangle)^2 \left. \frac{dG^2(n)}{dn^2} \right|_{\langle n \rangle} \right) + (1 - b_i) r,$$

where $G(n) = \frac{C_{V2R}}{n+1}$.

By taking the expectation of (2.14) on both sides with respect to n, we have

$$\langle R \rangle \approx b_i \left(G(\langle n \rangle) + \frac{1}{2} \mathrm{Var}(n) \left. \frac{dG^2(n)}{dn^2} \right|_{\langle n \rangle} \right) + (1 - b_i) \widetilde{\langle r \rangle}$$
$$\approx b_i C_{V2R} \left(\frac{1}{\langle n \rangle + 1} + \frac{\mathrm{Var}(n)}{(\langle n \rangle + 1)^3} \right) + (1 - b_2) \widetilde{\langle r \rangle}.$$

2.3.4.3 Mean File Download Delay

We evaluate the average file download delay by the mean first passage time stating from the state $K_i(0) = 0$, i.e., no blocks are downloaded, until the state $K_i(t) = \kappa_i$, i.e., the tagged node collects all the desired file blocks. Let $\Gamma(m, k)$ denote the first passage time starting when the vehicle is in state (m, k) until all κ_i blocks are downloaded, mathematically,

$$\Gamma_i(m, k) = \min\{t > 0 | M_i(0) = m, K_i(0) = k \text{ and } K_i(t) = \kappa_i\}.$$

The mean download delay of file i is thus

$$\tau_i = \frac{1}{\lambda + \mu} (\lambda \Gamma_i(0, 0) + \mu \Gamma_i(1, 0)), \tag{2.15}$$

where $\frac{\lambda}{\lambda+\mu}$ and $\frac{\mu}{\lambda+\mu}$ are the limiting probabilities that the tagged node is outside and inside the coverage of RSBs, respectively, when the tagged node initiates the download subscription of file i. The expression of τ_i is given in Theorem 2.2.

Theorem 2.2. *The average download delay of file i is*

$$\tau_i \approx \frac{b_i(\Phi-\delta)\lambda+\delta(\lambda+\mu)+\kappa_i(\lambda+\mu)^2}{b_i(\Phi-\delta)\mu(\lambda+\mu)+\delta(\lambda+\mu)^2}. \tag{2.16}$$

Proof. The mean first passage time from state $(0,k)$ can be represented in a recursive manner as

$$\Gamma(0,k) = \frac{1}{\mu+\delta} + \frac{\delta}{\mu+\delta}\Gamma(0,k+1) + \frac{\mu}{\mu+\delta}\Gamma(1,k), \tag{2.17a}$$

$$\Gamma(1,k) = \frac{1}{\lambda+\gamma} + \frac{\gamma}{\lambda+\gamma}\Gamma(1,k+1) + \frac{\lambda}{\lambda+\gamma}\Gamma(0,k), \tag{2.17b}$$

for $0 \leq k \leq \kappa_i - 1$, and

$$\Gamma(0,\kappa_i) = \Gamma(1,\kappa_i) = 0. \tag{2.18}$$

In (2.17a), the first term on the right-hand-side represents the mean time that the tagged node spends in state $(0,k)$. With probability $\frac{\delta}{\mu+\delta}$, the tagged node transits to state $(0,k+1)$ which has the mean first passage time $\Gamma(0,k+1)$; with the rest probability, it transits to state $(1,k)$ which has the mean first passage time $\Gamma(1,k)$. Equation (2.17b) is derived in the same manner.

As such, we have

$$\lambda\Gamma(0,k)+\mu\Gamma(1,k) \tag{2.19}$$

$$= \frac{A}{B} + \frac{\lambda+\mu}{B}(\lambda\delta\Gamma(0,k+1)+\mu\gamma\Gamma(1,k+1))$$

$$+ \frac{\delta\gamma}{B}(\lambda\Gamma(0,k+1)+\mu\Gamma(1,k+1)),$$

where $A = (\lambda+\mu)^2 + \lambda\gamma + \mu\delta, B = \lambda\delta + \gamma\mu + \delta\gamma$.

In particular, via (2.17a) we have

$$\lambda\delta\Gamma(0,k)+\mu\gamma\Gamma(1,k) = (\kappa_i-k)(\lambda+\mu). \tag{2.20}$$

By substituting (2.20) to (2.19), we have

$$\lambda\Gamma(0,k)+\mu\Gamma(1,k) \tag{2.21}$$

$$=\frac{A}{B}+\frac{(\kappa_i-k-1)(\lambda+\mu)^2}{\Pi}$$

$$+\frac{\delta\gamma}{B}(\lambda\Delta(0,k+1)+\mu\Delta(1,k+1))$$

$$=\cdots$$

$$=\sum_{i=0}^{\kappa_i-k-1}\left(\frac{\delta\gamma}{B}\right)^i\left(\frac{A}{B}+\frac{(\kappa_i-k-1)(\lambda+\mu)^2}{B}\right)$$

$$=\frac{1-\left(\frac{\delta\gamma}{B}\right)^{\kappa_i-k}}{1-\frac{\delta\gamma}{B}}\left(\frac{A}{B}+\frac{(\kappa_i-k-1)(\lambda+\mu)^2}{B}\right).$$

By plugging (2.21) into (2.15), we have

$$\tau_i=\frac{1}{\lambda+\mu}(\lambda\Gamma(0,0)+\mu\Gamma(1,0))$$

$$=\frac{1}{\lambda+\mu}\frac{1-\left(\frac{\delta\gamma}{B}\right)^{\kappa_i}}{1-\frac{\delta\gamma}{B}}\left(\frac{A}{B}+\frac{(\kappa_i-1)(\lambda+\mu)^2}{B}\right).$$

As $\frac{\delta\gamma}{B}<1$, when κ_i is large, we have $\left(\frac{\delta\gamma}{B}\right)^{\kappa_i}\approx0$, and accordingly,

$$\tau_i=\frac{1}{\lambda+\mu}\frac{1}{1-\frac{\delta\gamma}{B}}\left(\frac{A}{B}+\frac{(\kappa_i-1)(\lambda+\mu)^2}{B}\right)$$

$$=\frac{\gamma\lambda+\delta\mu+\kappa_i(\lambda+\mu)^2}{\gamma\mu(\lambda+\mu)+\lambda\delta(\lambda+\mu)} \tag{2.22}$$

By substituting (2.13), $\langle R\rangle=\gamma$ and $\langle r\rangle=\delta$ into (2.22), we have

$$\tau_i=\frac{b_i(\Phi-\delta)\lambda+\delta(\lambda+\mu)+\kappa_i(\lambda+\mu)^2}{b_i(\Phi-\delta)\mu(\lambda+\mu)+\delta(\lambda+\mu)^2}.$$

Corollary 2.1. *The download delay τ_i is a monotonic non-increasing convex function of b_i if $\frac{\kappa_i}{\delta}\geq\frac{1}{\mu}-\frac{2}{\lambda+\mu}$.*

Proof. According to Theorem 2.2, we obtain the first and second order derivative of τ_i as

$$\frac{d\tau_i}{db_i} = -\frac{(\Phi - \delta)\left[\kappa_i \mu (\lambda + \mu) + \delta (\mu - \lambda)\right]}{\left[b_i \mu (\Phi - \delta) + (\lambda + \mu)\right]^2}$$

$$\frac{d^2\tau_i}{db_i^2} = \frac{2\mu (\Phi - \delta)^2 \left[\kappa_i \mu (\lambda + \mu) + \delta (\mu - \lambda)\right]}{\left[b_i \mu (\Phi - \delta) + \delta (\lambda + \mu)\right]^3}$$

The download delay τ_i is a convex function if $\frac{d^2\tau_i}{db_i^2} \geq 0$, i.e., $\frac{\kappa_i}{\delta} \geq \frac{1}{\mu} - \frac{2}{\lambda + \mu}$.

2.4 Protocol Design of the RSB Infrastructure

By substituting (2.16) into (2.1), we are now ready to derive the solution of (2.8). We make two assumptions as follows:

- $\frac{\kappa_i}{\delta} \geq \frac{1}{\mu}$, for all i, which implies that the average download time of a file when vehicles are outside the RSBs (evaluated as κ_i/δ) should be no smaller than the average sojourn time of vehicle nodes outside the RSBs (evaluated as $1/\mu$). Otherwise, the assistance of RSBs is negligible as the desired file can be downloaded easily through V2V communications only before vehicles entering into the coverage of any RSBs.
- $U(\tau_i)$ is a non-increasing, twice differentiable concave function of τ_i. As an example to explain the methodology, we adopt

$$U(\tau_i) = -\tau_i \quad \text{and} \quad w_i = p_i, \ i \in \mathbf{F}, \tag{2.23}$$

for the simplicity.

According to Corollary 2.1 and Proposition 2.1, the network utility $\mathscr{U}$ is a concave function of b_i, and accordingly, the network utility maximization problem in (2.8) is a convex optimization problem.

Proposition 2.1. *If $U(\tau_i)$ is a non-increasing, twice differentiable concave function of τ_i, then $U(\tau_i)$ and network utility $\mathscr{U}$ are concave functions of b_i.*

Proof. Evaluating the first and second derivatives of $U(\tau_i)$ on b_i, we have

$$\frac{dU(\tau_i)}{dx_i} = \frac{dU(\tau_i)}{d\tau_i} \frac{d\tau_i}{db_i} \tag{2.24}$$

$$\frac{d^2U(\tau_i)}{db_i^2} = \frac{d^2U(\tau_i)}{d\tau_i^2} \left(\frac{d\tau_i}{db_i}\right)^2 + \frac{dU(\tau_i)}{d\tau_i} \frac{d^2\tau_i}{db_i^2} \tag{2.25}$$

Since $\frac{dU(\tau_i)}{d\tau_i} \leq 0$ and $\frac{d\tau_i}{db_i} \leq 0$, with (2.24a) we have $\frac{dU(\tau_i)}{db_i} \geq 0$. Since $\frac{d^2U(\tau_i)}{d\tau_i^2} \leq 0$, $\frac{dU(\tau_i)}{d\tau_i} \leq 0$ and $\frac{d^2\tau_i}{db_i^2} \geq 0$, with (2.24b), we have $\frac{d^2U(\tau_i)}{db_i^2} \leq 0$, and therefore, $U(\tau_i)$ is a concave function of b_i. As the network utility W, in general, is the weighted sum of the $U(\tau_i)$ over $i \in \mathbf{F}$, we have $\frac{d^2W}{db_i^2} = w_i \frac{d^2U(\tau_i)}{db_i^2} \leq 0$. Therefore, W is also a concave function of b_i.

2.4.1 Global Optimal Algorithm

Let $\mathbf{B}^*$ denote the optimal caching profile. By examining (2.8) with the Karush–Kuhn–Tucker (KKT) conditions as shown in Appendix 2.7.1, we have $\mathbf{B}^*$ as,

$$b_i^* = \frac{\sqrt{p_i[\kappa_i\mu(\lambda+\mu)+\delta(\mu-\lambda)]}}{\sum_{j\in\mathbf{F}}\kappa_j\sqrt{p_j[\kappa_j\mu(\lambda+\mu)+\delta(\mu-\lambda)]}}\left(\mathscr{L}+\frac{\delta(\lambda+\mu)}{\mu(\Phi-\delta)}\sum_{j\in\mathbf{F}}\kappa_j\right) - \frac{\delta(\lambda+\mu)}{\mu(\Phi-\delta)}, \quad b_i^* \in \mathbf{B}^*.$$

$$(2.26)$$

Then given the availability profile $\mathbf{A}$ and popularity profile $\mathbf{P}$, each RSB should select content file i to cache with probability b_i^*. We refer to this scheme as the global optimal replication in RSBs.

The global optimal replication provides the optimal solution). However, note that both $\mathbf{A}$ and $\mathbf{P}$ are system-wide parameters which relate to the file information across the whole network. They are not available to individual RSBs or vehicle nodes when the network size is large. Therefore, the global optimal replication scheme is not practical for large-scale real-world deployment. Nevertheless, the global optimal replication provides a benchmark for performance comparison with other replication schemes. In what follows, we present a decentralized algorithm to determine the content replication at RSBs.

2.4.2 Distributed Content Replication

In this part, we design a distributed algorithm to enable RSBs to select the appropriate files to store according to (2.8) in a fully distributed manner. To achieve this goal, we approximate b_i^* by b_i^d as

$$b_i^d = \frac{\mathscr{L}\sqrt{p_i[\kappa_i\mu(\lambda+\mu)+\delta(\mu-\lambda)]}}{\sum_{j\in\mathbf{F}}\kappa_j\sqrt{p_j[\kappa_j\mu(\lambda+\mu)+\delta(\mu-\lambda)]}}. \tag{2.27}$$

This can greatly simplify the algorithm design with modest performance degradation as verified by simulations.

To help RSBs distributedly select file i with the probability b_i^d from the network, we adopt a random walk based algorithm over a file graph as follows:

2.4.2.1   File Graph

The file graph refers to as a graph connecting all the files stored in distributed vehicles. As an example shown in Figure 2.3, each vertex in the graph represents a file stored in a vehicle node. Additionally, each vehicle has an anchor file, e.g., file j, which is selected from the locally stored files in vehicles and has the largest value of $\sqrt{p_j \Phi \left[\kappa_j \mu (\lambda + \mu) + \delta (\mu - \lambda) \right]}$ among the buffered files. Each vehicle node periodically broadcasts its anchor file information, including the availability a_j and popularity p_j, to the neighbor vehicles. How to measure the availability and download demand of files will be described in Sect. 2.4.2.2. In the file graph, all files stored in the same vehicle node are fully connected, and the anchor files among neighboring vehicles are fully connected, as shown in Figure 2.3. Therefore, the file graph has a two-tier architecture where the top tier connects the anchor files of vehicles and the underlying tier connects all the files inside a vehicle to its anchor file.

2.4.2.2   Random Walk Based File Selection

The file selection is realized by a random walk algorithm over the file graph as described in Algorithm 1. Specifically, to determine the files stored in RSBs, an RSB first issues a number η of random walkers to separate vehicles in the communication range. Each vehicle which receives a walker will then initiate the random walk process starting from its anchor file. The walker is forwarded stochastically on the file graph from one vertex (file) to another vertex (file) following the Metropolis–Hasting algorithm; the derivation of transition probabilities in the random walk is given in Appendix 2.7.2. Once the walker is forwarded to the anchor file, it may be relayed to other anchor files stored in different vehicles. In this case, the walker is forwarded to other vehicles and proceeds the random walk algorithm. After being relayed for Time-To-Live (TTL) hops among files on the file graph including self-loops, the walker stops at a file which is then selected to be uploaded to RSBs.

In order to compute the transition probability of the walker, each vehicle needs to know the availability and popularity of the files stored in its buffer. In the RSB infrastructure, we enable vehicles to distributedly measure these parameters as follows:

Measurement of File Availability

The availability a_i of file i is only measured by the vehicles which need to download file i. As each vehicle interested in file i continually issues the download requests to its neighboring vehicles, it can estimate the file availability a_i based on the replies with $\frac{\text{No. of vehicles having file } i \text{ stored}}{\text{Overall No. of vehicles contacted}}$. Whenever the vehicle, e.g., x, interested in file i meets another vehicle, e.g., y, which has file i stored, vehicle x would inform the

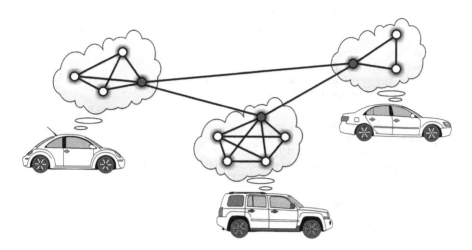

Fig. 2.3 File graph in the vehicular network

measurement of a_i to vehicle y piggybacked with the download request. As such, vehicle y would receive multiple measurements of a_i. For each new measurement received, it would incorporate it with the previous measurement using the moving average. Once vehicle x finishes downloading file i, it can use its measurement on a_i to evaluate the availability of file i.

Measurement of File Popularity

The download demand d_i of file i is measured by the vehicles which have file i stored. As those vehicles keep receiving download requests from others and a portion of the requests are for file i, d_i can be estimated based on this information as $\frac{\text{No. of vehicles requesting to download file } i}{\text{Overall No. of download requests received}}$.

It is important to note that each vehicle only needs to know the available and download demand of the files it stores. Therefore, the distributed measurement will not impose much workload on the message exchange. To improve the measurements of availability and download demand of files, we can also make use of the RSBs. In this case, RSBs would collect the measurements from different vehicles driving through, and average the measurements towards a more accurate estimation, then announce them to vehicles. There would be other methods for more accurate measurements based on the vehicular sensor networks [12], which is out of the scope of this work.

Algorithm 1: Random walk algorithm starting from file x

```
/* m: current file with walker                              */
/* h: hop account                                            */
/* p: random number                                          */
/* Pmn: transition probability from file m to file n shown in
       (2.39) of Appendix 2.7.2                              */
```

begin

    Initialization: $m \leftarrow x$; $h \leftarrow 0$; $p \leftarrow 0$;

    while $h < TTL$ **do**

        $p \leftarrow$ random number in $[0, 1]$;

        foreach *file n (n $\neq$ m) connected to file m in the file graph* **do**

            if $p \leq P_{mn}$ **then**

                $m \leftarrow n$;

                quit the **foreach** loop;

            else

                $p \leftarrow p - P_{mn}$;

        $h \leftarrow h + 1$;

Result: File m

2.4.3 Protocol Description

This part describes the detailed protocol design and implementation of the RSB infrastructure. In the network, each RSB, e.g., A, works in a fully distributed manner and conducts the following three operations:

2.4.3.1 File Publication

Whenever a new file is published at RSB A (uploaded by its owner), RSB A issues η walkers to separate vehicles in its coverage. Each walker is relayed among files over the file graph embedded in the vehicular networks following Algorithm 1, and results in one file selected after the TTL hops. The vehicles with the selected files will then upload the files to the RSBs which they drive through. As such, RSBs are dynamically refreshed with new contents continuously uploaded; and this process is triggered by the publication of new files. The value of η will be discussed later. Note that in this phase, RSB A is only responsible to issue walkers to the vehicular network upon the publication of new files. The files selected by the walkers will be uploaded to RSBs in the communication range of the vehicles hosting the selected file, which may not be RSB A.

    The value of η is set to make the overall number of files in the network stable. It is dependent on the rates at which new files are published to the network and the outdated files are evicted from the network. Let $\mathscr{T}$ be the average life time of files in the network, where the life time represents the time duration that a file is stored in RSBs.

Let θ be the average injection rate of new contents to the network at distributed RSBs. As each newly published file will initiate η walkers to the vehicular network and finally cause η files to be uploaded to RSBs from vehicles, the rate at which RSBs get new contents uploaded is in total $(1+\eta)\theta$. Let $\mathcal{N}$ denote the number of RSBs in the overall network. Mathematically, the rate at which the number of content files changes over time is

$$\frac{\partial |\mathbf{F}|}{\partial t} = (1+\eta)\theta\mathcal{N} - \frac{1}{\mathcal{T}}\mathcal{N}\mathcal{L}. \tag{2.28}$$

In the steady state with $\frac{\partial |\mathbf{F}|}{\partial t} = 0$, we have

$$\eta = \frac{\mathcal{L}}{\mathcal{T}\theta} - 1. \tag{2.29}$$

To compute η with (2.29), we assume that θ and $\mathcal{T}$ are known through measurements at different RSBs distributively based on the history of file storage in RSBs. RSBs can also exchange the measurements among each other to improve the accuracy with the assistance of vehicles.

2.4.3.2 Retrieve Files from Vehicles

Whenever a vehicle with a selected file in the random walk algorithm comes into the coverage of RSB A, it will retrieve the file immediately from the vehicle. During this period, the channel of RSB A is used exclusively for the file retrieval. If there are multiple uploads simultaneously from different vehicles to RSB A, RSB A only processes one retrieval at one time until this retrieval completes. Once the selected file in a vehicle is uploaded, the vehicle will not upload this file to other RSBs unless this file is selected again in the random walk algorithm. In case that a vehicle moves out of the coverage of RSB A before it accomplishes the retrieval, RSB i would proceed the file retrieval again from other driving through vehicles which has the unfinished file stored. If its buffer is full, RSB A depletes the buffer by deleting the file which has been stored for the longest time.

2.4.3.3 Upload File to Vehicles

In the idle period of RSB A when it does not need to issue walkers to the network or retrieve files from vehicles, it uploads the cached file to the driving through vehicles upon their requests.

Each RSB in the network thus works in the three modes in a fully distributed manner. In what follows, we evaluate the performance of the RSB infrastructure compared to the centralized content replication.

2.5   Simulation Verification

This section evaluates the performance of the RSB infrastructure using simulations based on a discrete-event simulator coded in C++.

2.5.1   Simulation Setup

Our simulation is carried out over a 1.5×1.5 km regional road map on the Manhattan island with the contour of the street layout plotted in Figure 2.4. Each road segment in Figure 2.4a is of two lanes with the bidirectional vehicle traffic. Compromised to the complexity of simulations, we select a bounded region on the map for our simulations, as shown in Figure 2.4b. There are totally 29 RSBs deployed in the region with the communication range uniformly distributed within the range $[180, 200]$ m. For each simulation run, 300 vehicles are involved in the content distribution. The mobilities of vehicles are generated by VANETMobisim [14], in which the destination of each trip is randomly selected, and the velocity of each vehicle is controlled no larger than 60 km/s and adapted by the IDM-LC(Intelligent Driver Model with Lane Changes) mode. The coverage of V2V communications is set to be 150 m. With this configuration, we have $\lambda = 29.46$ s, $\mu = 12.19$ s, $\langle n \rangle = 4.02$ and $\mathrm{Var}(n) = 6.18$. In each simulation run, 200 files are initially available for download in the network, which are randomly stored in RSBs and vehicles. The file size is accounted in the unit of blocks and uniformly distributed within $[40, 100]$ blocks. Unless mentioned otherwise, all RSBs have the equal buffer storage to cache 3×10^3 file blocks, i.e., $3 \times 10^3 / 100 = 30$ files at most. Vehicles have equal

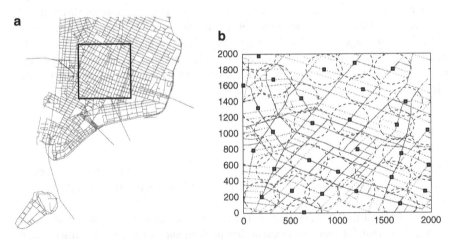

Fig. 2.4 Street layout based TIGER/Line Shapefiles [13], and RSB distribution in simulations (**a**) Street layout (**b**) RSB distribution on the road

Fig. 2.5 Mean download rate
with different values of file
cache probability b

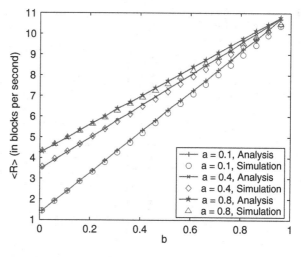

Fig. 2.6 Mean file download
delay with different values of
file cache probability b

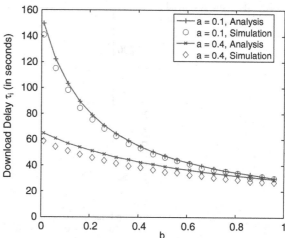

buffer to cache 1×10^3 file blocks, i.e., 10 files at most. The download capacity
of the vehicle to RSB communication, C_{V2R}, is 50 blocks/s and that of the V2V
communication, C_{V2V}, is 20 blocks/s. Each vehicle can communicate with one other
network component at most, and the parallel communication sessions are scheduled
through the ideal MAC.

2.5.2   Verification of the Analysis

As the presented protocol is based on the evaluation of file download delay, in the
first experiment, we verify the accuracy of (2.10), (2.13) in evaluating the mean

Fig. 2.7 Mean download rate
with different values of file
availability a

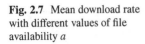

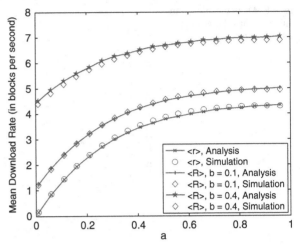

Fig. 2.7 Mean download rate with different values of file availability a

Fig. 2.8 Mean file download
delay with different values of
file availability a

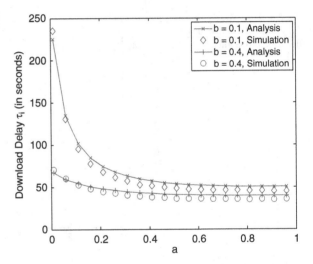

download rates $\langle r \rangle$ and $\langle R \rangle$, when vehicles are inside and outside RSBs, respectively, and the accuracy of (2.16) in evaluating the mean download delay of files. To this end, we carry out the Mento Carlo simulations by investigating on the download performance of a tagged vehicle. We make the tagged vehicle subscribe to download a file, referred to as file i in this section, of file size to be 100 blocks and report the averaged results over 5,000 simulation runs.

Figure 2.5 shows the values of $\langle r \rangle$ and $\langle R \rangle$ as a function of b_i when a_i is 0.1. As we can see from the figure, $\langle r \rangle$ remains the same with different b_i, and $\langle R \rangle$ increases linearly with b_i. The analyses in (2.10), (2.13) match the simulations well. Figure 2.6 shows the mean download delay of the file with different b_i. As we can see, when b_i increases, the download delay τ_i reduces dramatically which can be

Fig. 2.9 Mean download rate with different values of $\langle n \rangle$

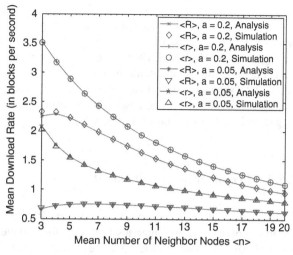

Fig. 2.10 Mean file download delay with different values of $\langle n \rangle$

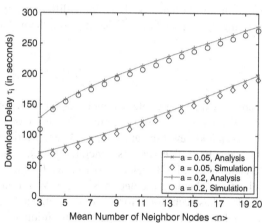

characterized by (2.16). Moreover, τ_i is a convex function of b_i which validates Corollary 2.1. In addition, when a_i changes from 0.1 to 0.4, the download delay τ_i reduces significantly, as in this case more vehicles on the road have file i stored in their local buffer and therefore the tagged node can finish downloading faster.

Figure 2.7 shows the values of $\langle r \rangle$ and $\langle R \rangle$ as a function of a_i when b_i is 0.1. As we can see, by increasing a_i, both $\langle r \rangle$ and $\langle R \rangle$ increase with a constant gap between the two curves which can be characterized by (2.13). As shown in Figure 2.8 and indicated by (2.16), the mean download delay of file i reduces when a_i increases.

Figure 2.9 shows the values of $\langle r \rangle$ and $\langle R \rangle$ as a function of the mean number of neighbor vehicles, i.e., $\langle n \rangle$, when b_i is 0.1. As we can see, by increasing $\langle n \rangle$, $\langle R \rangle$ reduces monotonically. This is because that with $\langle n \rangle$ increasing more vehicles share the capacity of RSBs and contend the channel with the tagged node. As in

Figure 2.9, in both cases when $a_i = 0.2$ and 0.05, respectively, $\langle r \rangle$ increases first when $\langle n \rangle$ increases and then reduces. This is because that when $\langle n \rangle$ increases, more neighbor vehicles may have the desired file i stored and upload the file to the tagged node. However, when $\langle n \rangle$ is large, indicating that more neighbor nodes are contending the channel with the tagged node, $\langle r \rangle$ reduces with $\langle n \rangle$ increasing. Figure 2.10 shows the download delay of file i, τ_i, as a function of $\langle n \rangle$. As we can see, by increasing $\langle n \rangle$, τ_i increases. This is because that the download rate of the tagged node reduces as shown in Figure 2.9.

2.5.3 Performance of Protocol

We simulate a dynamic network in which each RSB periodically publishes a new file to the network at the intervals following the exponential distribution with the mean of 60 s. The index of files increases linearly according to the publication time of the file in the network. Files have different popularity which follows the Zipf distribution; the popularity of the ith file in the network is as

$$p_i = \frac{1}{(\hat{i})^\alpha} / \sum_{j=1}^{|\mathbf{F}|} \frac{1}{j^\alpha}, \qquad (2.30)$$

where α is a configurable parameter of the Zipf function. $\hat{i} = (i \bmod 500)$ where mod denotes the modulo operation. In this case, the popularity of files renews whenever 500 new files are published. The lifetime of each file is set to be 200 s which results in five walkers generated per RSB according to (2.29) when a new file is published. RSBs selectively retrieve files from vehicles based on the content replication scheme presented in Sect. 2.4. When the buffers of RSBs overflow, the file which has been stored for the longest time in the buffer is evicted. Vehicles select files to download based on the Zipf distribution as aforementioned. Once the buffer of vehicles is full, a randomly selected file is evicted to release the cache for new downloads.

We evaluate the utility function $\mathscr{U} = -\sum_{i \in \mathbf{F}} p_i \tau_i$ every 100 sections, which accounts for the summed download delay of files weighted by the file popularity within this period. We conduct 50 runs upon each simulation experiment and plot the mean result with the 95 % confidence intervals.

Figure 2.11 shows the comparison between the presented random walk based content replication scheme and the global optimal and local greedy content replication schemes. The three schemes adopt the same content upload and download operations between vehicles and RSBs, except for the file selection strategies when individual RSBs retrieve files to store. Using the global optimal scheme, each RSB selects a file, e.g., i, from the drive through vehicles to store with the probability b_i^* as shown in (2.26). Using the local greedy algorithm, each RSB selects a file with the largest value of file popularity p_i to store. The

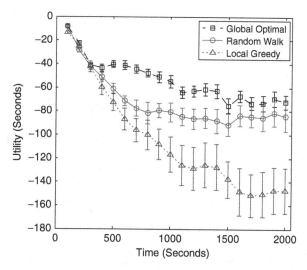

Fig. 2.11 Comparison between global optimal, random walk based and local greedy content replication schemes

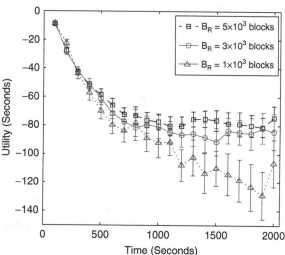

Fig. 2.12 Global network utility with different B_R

strategy of file selection using the random walk based algorithm is discussed in Sect. 2.4. As we can see from Figure 2.11, the global optimal scheme has the best performance, followed by the random walk algorithm. The local greedy algorithm has the worst performance for two reasons. Firstly, by selecting files with the local maximal file popularity to store, the local greedy scheme is myopic and cannot optimize the overall performance of the network. Secondly, without considering the storage of vehicles, using the local greedy scheme, RSBs may store files which have already been vastly stored in vehicles, and therefore, cannot be efficiently utilized to cooperate with the vehicular storage towards maximal social welfare. Notably, in all the schemes, the global network utility $\mathcal{U}$ reduces over time and finally approaches

Fig. 2.13 Global network
utility with different C_{V2R}

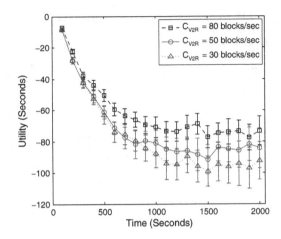

Fig. 2.14 Global network
utility with different vehicular
buffer B_{veh}

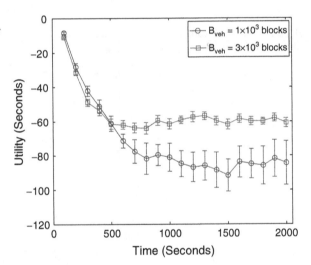

to a stabilized value. This is because that at each interval, $\mathcal{U}$ is evaluated by summing up the weighted download delays of files downloaded. Therefore, in the early period of simulations, only files with small delays are finished and accounted, leading to the small value of $\mathcal{U}$. Eventually, files with long download delays are accounted when they are finished, and the value of $\mathcal{U}$ becomes stable accordingly.

Figures 2.12 and 2.13 show the global network utility with different buffer sizes and communication capacity of RSBs, respectively, with other parameters remaining the same. As we can see, when the buffer size or communication capacity increase, the global network utility increases, indicating smaller download delay of files. However, enhancing the buffer storage and communication capacity will lead to the increased physical cost of RSBs which discourages the large-scale deployment of RSBs.

Fig. 2.15 Global network utility with different C_{V2V}

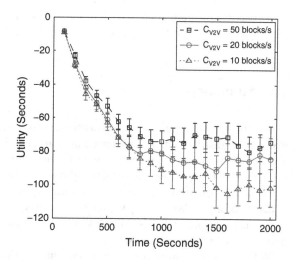

Fig. 2.16 Performance with different values of α

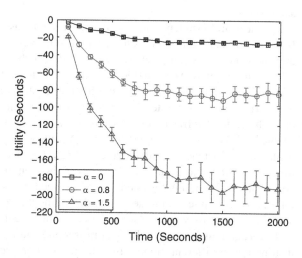

Figures 2.14 and 2.15 show the global network utility when increasing the buffer size and V2V capacity of vehicles, respectively, with other parameters remaining the same. As we can see, increasing the vehicles' buffer size or the capacity of V2V communications will significantly improve the network performance, as more bandwidth and storage resource are available in the network. However, in practice, the capacity and buffer storage contributed by vehicles are out of control. This therefore calls for an effective incentive mechanism to encourage contributions.

Figure 2.16 shows the global network utility with different values of α in the Zipf function (2.30). $\alpha = 0$ indicates that all the files have the equal popularity; the larger α is, the faster that the popularity of files decreases when $\hat{i}$ increases. When α increases, the global network utility reduces significantly. This is because that there exists certain very popular files which are highly demanded. As such, the

RSBs become the upload bottleneck of the files which enlarge the download delay of vehicles. To address this effect requires to increase the upload capacity of RSBs.

2.6 Summary

In this chapter, we have described a distributed large-scale infrastructure for vehicular content distribution. The presented infrastructure is formed by RSBs deployed across the city which are managed by individual entities at different locations and form an integrated system towards the global network utility. To enable RSBs to work in a fully distributed manner and accordingly make the presented infrastructure scalable to any network size, we have introduced a random-walk based content replication scheme at RSBs. With extensive simulations, we have validated the performance of the presented scheme.

Within this framework described in the chapter, there exist multiple interesting and open issues:

Connection to Internet: in this chapter, RSBs are assumed to be unconnected to Internet. This can avoid the expensive bandwidth cost to RSB owners and encourage them to deploy RSBs. In practice, certain RSBs which are connected to the Internet, referred to as Internet-enabled RSBs, can be strategically deployed. In this case, by relying on vehicles to retrieve the Internet contents and transport the contents to different locations, limited Internet content distribution services can be provided through the RSB infrastructure. In this paradigm, the locations of these Internet-enabled RSBs are crucial to reduce the download delay of Internet contents to vehicles. For instance, the deployment of Internet-enabled RSBs should take the trajectories of vehicles and the density of the non Internet-enabled RSBs into consideration. From this perspective, [15] has shed important lights by investigating on the content replication issue, i.e., to determine the optimal locations of contents stored in the deployed infrastructure based on the trajectories of vehicles.

Heterogenous Users: the RSB infrastructure can be extend to provide content distribution to different mobile users in different environments. For example, the RSBs can be deployed in a shopping mall to distribute store flyers to users with tablets, PDAs and laptops. In this case, different users may have different characteristics of mobility and requirements on the service quality. This dictates the network to take the distinct features of heterogenous user's QoS requirements into considerations.

Security Threat: without central control, the RSB infrastructure faces multiple security threats. For example, the buffer storage of RSBs can be abused to store and distribute harmful contents or virus to vehicles. Moreover, the contents stored in RSBs may also be polluted by garbage contents with misleading and mismatched titles. The content population is severe and has been extensively investigated in peer-to-peer networks [16], which however has not been addressed in the vehicular content distribution networks. To combat the security issues, it is necessary for RSBs to quickly identify and filter the harmful and spam contents [17].

2.7 Appendix

2.7.1 Optimal Solution of (2.8) According to KKT Conditions

As (2.8) is a convex optimization problem, the KKT conditions are both necessary and sufficient for the optimal solution. Let b_i^* denote the optimal solution of (2.8). Introducing Lagrangian multiplier ϖ for the constraint in (2.8), we list the KKT conditions of (2.8) as follows:

$$\left.\frac{\partial \mathscr{U}}{\partial b_i}\right|_{b_i^*} - \varpi \kappa_i = 0 \tag{2.31}$$

$$\varpi \left(\mathscr{L} - \kappa_i b_i^* \right) = 0 \tag{2.32}$$

$$\sum_{i \in \mathbf{F}} \kappa_i b_i^* \leq \mathscr{L} \tag{2.33}$$

$$b_i^*, \gamma \geq 0,\, i \in \mathbf{F} \tag{2.34}$$

where $\mathscr{U} = -\sum_{i \in \mathbf{F}} p_i \tau_i$ as specified in (2.1) and (2.23)

Assume that $\varpi = 0$. Substituting it into (2.31), we have $p_i \left.\frac{\partial \tau_i}{\partial b_i}\right|_{b_i^*} = 0$. This is not feasible as $\frac{\partial \tau_i}{\partial b_i} < 0$ and $p_i > 0$ for all $i \in \mathbf{F}$. Therefore, from (2.32) we have $\gamma > 0$ and $\mathscr{L} - \sum_{i \in \mathbf{F}} b_i^* = 0$.

Substituting (2.16) into (2.31), we have

$$b_i^* = \frac{1}{\mu(\Phi - \delta)} \sqrt{\frac{p_i(\Phi - \delta)\left[\kappa_i \mu (\lambda + \mu) + \delta(\mu - \lambda)\right]}{\varpi \kappa_i}} - \frac{\delta(\lambda + \mu)}{\mu(\Phi - \delta)}. \tag{2.35}$$

Together with $\mathscr{L} - \sum_{i \in \mathbf{F}} b_i^* = 0$, we have

$$b_i^* = \mathscr{L} \frac{\sqrt{p_i \left[\kappa_i \mu (\lambda + \mu) + \delta(\mu - \lambda)\right]}}{\sum_{j \in \mathbf{F}} \kappa_j \sqrt{p_j \left[\kappa_j \mu (\lambda + \mu) + \delta(\mu - \lambda)\right]}}$$
$$+ \left(\frac{\sqrt{p_i \left[\kappa_i \mu (\lambda + \mu) + \delta(\mu - \lambda)\right]} \sum_{i \in \mathbf{F}} \kappa_i}{\sum_{j \in \mathbf{F}} \kappa_j \sqrt{p_j \left[\kappa_j \mu (\lambda + \mu) + \delta(\mu - \lambda)\right]}} - 1 \right) \cdot \frac{\delta(\lambda + \mu)}{\mu(\Phi - \delta)}, \tag{2.36}$$

and

$$\varpi = \left(\frac{\sum_{i \in \mathbf{F}} \sqrt{\kappa_i p_i \left[\kappa_i \mu (\lambda + \mu) + \delta(\mu - \lambda)\right]}}{\mathscr{L}\mu + \delta(\lambda + \mu)\sum_{i \in \mathbf{F}} \kappa_i} \right)^2$$

2.7.2 Transition Probability in the Random Walk Algorithm

The target of the random walk algorithm is to select a file i from the file graph with the probability b_i^d shown in (2.27).

Assume that there are totally V nodes presenting in the network and a_i of them having file i stored. Therefore, there are totally $a_i |V|$ copies of file i in the file graph. To select file i with probability b_i^d, one should sample each copy of file i in the graph with probability

$$\pi_i = \frac{b_i^d}{a_i V}. \tag{2.37}$$

Using the Metropolis–Hasting algorithm, the transition of random walk constitutes two steps. In the first step, a candidate file, e.g., m, is selected from the neighboring files of the current file, e.g., n, which holds the walker based on the proposal probability

$$\alpha_{mn} = \frac{1}{s_m + 1}, \tag{2.38}$$

where s_m denotes the fanout of file m in the file graph. A neighboring file of file n is the file which is connected to file n in the file graph.

In the second step, file m is accepted as the next hop of the walker with the acceptance probability as

$$q_{mn} = \min\left\{ \frac{\pi_n \alpha_{nm}}{\pi_m \alpha_{mn}}, 1 \right\} = \min\left\{ \frac{a_m b_n^d (s_m + 1)}{a_n b_m^d (s_n + 1)}, 1 \right\},$$

with the rest probability the walk will sojourn in file n for one hop.

Therefore, the transition probability from file m to file n is

$$P_{mn} = \alpha_{mn} q_{mn} = \begin{cases} \frac{1}{s_m+1} \min\left\{ \frac{a_m b_n^d (s_m+1)}{a_n b_m^d (s_n+1)}, 1 \right\}, & m \neq n, \\ 1 - \sum_{m \neq n} P_{mn}, & m = n. \end{cases} \tag{2.39}$$

References

1. N. Post, "Toronto Hydro Assailed for City-Wide WiFi Plan." Canada.com, March 7 2006.
2. M. Reardon, "Cities Deploying Wi-Fi Face Challenges." CNET News, May 1 2006.
3. D. Hadaller, S. Keshav, T. Brecht, and S. Agarwal, "Vehicular Opportunistic Communication Under the Microscope," in *Proc. of ACM MobiSys*, 2007.
4. T. H. Luan, X. Ling, and X. Shen, "MAC in Motion: Impact of Mobility on the MAC of Drive-Thru Internet," *IEEE Transactions on Mobile Computing*, vol. 11, no. 2, pp. 305–319, 2011.

5. M. Asefi, J. W. Mark, and X. Shen, "A Mobility-Aware and Quality-Driven Retransmission Limit Adaptation Scheme for Video Streaming over VANETs," *IEEE Transactions on Wireless Communications*, vol. 11, no. 5, pp. 1817–1827, 2012.

6. B. Yu and F. Bai, "ETP: Encounter Transfer Protocol for Opportunistic Vehicle Communication," in *Proc. of IEEE Infocom*, 2011.

7. T. H. Luan, L. X. Cai, J. Chen, X. Shen, and F. Bai, "Engineering a Distributed Infrastructure for Large-Scale Cost-Effective Content Dissemination over Urban Vehicular Networks," in *IEEE Transaction on Vehicular Technology*. vol. 63, no. 3, pp. 1419–1435, 2013.

8. F. Bai and B. Krishnamachari, "Spatio-Temporal Variations of Vehicle Traffic in VANETs: Facts and Implications," in *Proc. of ACM VANETs*, 2009.

9. Y. Zhuang, J. Pan, Y. Luo, and L. Cai, "Time and Location-Critical Emergency Message Dissemination for Vehicular Ad-Hoc Networks," *IEEE Journal on Selected Areas in Communications*, vol. 29, pp. 187–196, Jan. 2011.

10. A. Nandan, S. Das, G. Pau, M. Gerla, and M. Y. Sanadidi, "Co-Operative Downloading in Vehicular Ad-Hoc Wireless Networks," in *Proc. of IEEE/IFIP WONS*, 2005.

11. F. R. K. Chung and L. Lu, *Complex Graphs and Networks*. Amer Mathematical Society, 2006.

12. U. Lee, E. Magistretti, M. Gerla, P. Bellavista, and A. Corradi, "Dissemination and Harvesting of Urban Data Using Vehicular Sensing Platforms," *IEEE Transactions on Vehicular Technology*, vol. 58, pp. 882–901, Feb. 2009.

13. "U.S. Census Bureau TIGER/Line Shapefiles & Files," Retrieved in Feb. 2012.

14. J. Haerri, F. Filali, C. Bonnet, and M. Fiore, "VanetMobiSim: Generating Realistic Mobility Patterns for VANETs," in *Proc. of ACM VANET*, 2006.

15. Y. Huang, Y. Gao, K. Nahrstedt, and W. He, "Optimizing File Retrieval in Delay-Tolerant Content Distribution Community," in *Proc. of IEEE ICDCS*, 2009.

16. J. Liang, R. Kumar, Y. Xi, and K. W. Ross, "Pollution in P2P File Sharing Systems," in *Proc. of IEEE Infocom*, 2005.

17. R. Lu, X. Lin, T. H. Luan, X. Liang, X. Li, L. Chen, and X. Shen, "PReFilter: An Efficient Privacy-preserving Relay Filtering Scheme for Delay Tolerant Networks," in *Proc. of IEEE Infocom*, 2012.

Chapter 3
Medium-Size Content Transmission Over Infrastructure-Less Inter-Vehicle Communications

In this chapter, we discuss on the content distribution over highly dynamic inter-vehicle or V2V communications. In particular, we consider the transmission of a single content file between a pair of vehicles, as shown in Figure 3.1, which represents a fundamental issue to the upper-layer content distribution services of distributing medium-size files from one vehicle to multiple vehicles.

3.1 Introduction

A basic requirement in content distribution is that the content files need to transmitted in their entirety from the sender to the reliever. Contents which are only partially transmitted, referred to as the fragment contents in this chapter, are typically unusable as they can hardly be successfully presented by the upper-layer applications. In real-world deployments, fragment contents are mainly caused by the connection interruptions. They are not only annoying to the end users with the deferred response or even failed presentation of the application, but also the transmission of fragment contents severely waste the precious network bandwidth by injecting invalid data to the network. Therefore, to avoid the fragment content transmissions represent a fundamental issue in content distribution networks.

The issue of fragment contents tends to be even severe in V2V-based vehicular content transmissions. This mainly attributes to the dynamic nature of the V2V connections, which can be described from the following three aspects:

- *Transient Connectivity*: with the diverse velocities of vehicles, the distance between vehicles is dramatically changing over time, resulting in the short-lived and intermittent V2V connectivity. For example, it is observed that the average connection time among vehicles on the highway in a real-world measurement is only around 15 s [1].

T.H. Luan et al., *Enabling Content Distribution in Vehicular Ad Hoc Networks*, SpringerBriefs in Computer Science, DOI 10.1007/978-1-4939-0691-8_3,

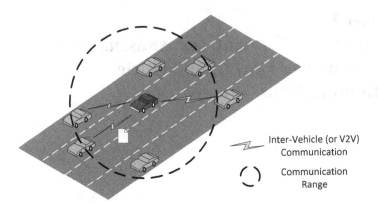

Fig. 3.1 Inter-vehicle communications on the highway

- *Harsh Wireless Channel*: due to the high mobility, severe Doppler shift and absence of line-of-sight communications,[1] the V2V channel suffers from severe fading and channel impairments. As reported in [3], the throughput of V2V communications in a real-world measurement is less than one fifth of the throughput of the vehicle-to-infrastructure communication.
- *Intense Channel Contention*: the vehicular network is typically large-scale with a multitude of vehicles sharing (or contending) the channel simultaneously. For example, as indicated in [4] a stable highway traffic flow typically constitutes 20–30 vehicles per mile per lane. In other words, in an eight-lane bidirectional highway section with smooth traffic flow and V2V communication range to be around 300 m, approximately 30–45 vehicles will share the channel for transmissions at the same time.

Note that a typical content file, such as a MP3 or video clip, is of several MBs, which may take tens of seconds or several minutes to transmit at the rate of several tens or hundreds Kbps. Over the short-lived and spotty V2V channel as discussed above, content transmissions may terminate with contents only partially transmitted. As the content can hardly be accomplished if without a careful design, leading to poor system performance with rampant fragment contents and invalid transmissions.

In this chapter, we describe an integrity-oriented framework presented in [5] to enable the reliable content transmissions over the dynamic V2V communications. We unfold our journey in three steps. Firstly, we focus on the content transmissions between a randomly selected source-destination pair of vehicles on a highway road scenario, and develop an analytical model to evaluate the data transmission performance in terms of the data volume that can be transmitted during the transient connection time of selected vehicles. Our analysis comprises of three components

---

[1]With the heterogenous heights and shapes of vehicles, the line-of-sight link could be blocked by intermediate vehicles as obstacles on the highway [2].

on the modeling of headway distance between vehicles, wireless channel fading and MAC contentions, respectively, and represent the data transmission performance as a function of the initial headway distance between the source-destination pair and the statistics of velocities. Based on the developed model, we are able to evaluate the likelihood of successful content transmission during the short-lived V2V connection. After that, we describe an adaptive admission control scheme to manage the content transmission subscriptions among vehicles. Using described framework, the file transmissions which can hardly be finished during the short-lived V2V connections will be rejected before the transmission commences. This thus prevents the potential bandwidth waste and invalid transmissions. Lastly, using extensive simulations, we show the accuracy of the developed model and validate the effectiveness of the described admission control protocol.

3.2 Model of the Content Transmission Over V2V Connections

We investigate on the highway vehicular network, where all the vehicles move over a linear topology. In particular, we focus on the content transmission between a randomly selected pair of vehicles, namely i and j. Our goal is to evaluate the impacts of network dynamics, from the aspects of transient connection time, channel fading and MAC contentions, on the data transmission performance from i to j in terms of the data volume that can be transmitted along the short-lived, spotty V2V channel. Our analysis is proceeded in four steps as following:

Step 1: *Headway distance prediction* to estimate $\mathcal{H}_{ij}(t)$ which denotes the headway distance from i to j after duration t starting when the evaluation is initiated.

Step 2: *Channel fading analysis* to evaluate the physical layer capacity, denoted by $\mathcal{C}_{ij}$, between i and j. At a given time, $\mathcal{C}_{ij}$ is a function of the headway distance $\mathcal{H}_{ij}(t)$ and is subject to the fast channel fading.

Step 3: *Channel contention model* to derive the effective transmission rate from i to j, denoted by $\mathcal{R}_{ij}$, at the MAC layer after the channel contentions. $\mathcal{R}_{ij}$ is the MAC throughput dependent on the underlying physical layer transmission rate $\mathcal{C}_{ij}$ and number of vehicles sharing the channel.

Step 4: *Download performance evaluation* to evaluate the data volume, denoted by $\mathcal{A}_{ij}$, that can be transmitted from i to j by integrating the MAC throughput $\mathcal{R}_{ij}$ over the V2V connection time.

In what follows, we derive the expressions of $\mathcal{H}_{ij}(t)$, $\mathcal{C}_{ij}$, $\mathcal{R}_{ij}$ in sequence and finally evaluate $\mathcal{A}_{ij}$. As $\mathcal{H}_{ij}(t)$ is a function of time, $\mathcal{C}_{ij}$ and $\mathcal{R}_{ij}$ also change over time. In brief of notation, we drop the subscript ij in the rest part of this section.

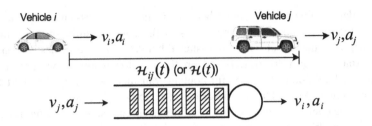

Fig. 3.2 Model of headway distance from i to j

3.2.1 Headway Distance Prediction

We consider the headway distance $\mathscr{H}(t)$ as a directional variable from the source vehicle i to the destination vehicle j, as shown in Figure 3.2. $\mathscr{H}(t) \geq 0$ if vehicle i is behind vehicle j in the moving direction; otherwise, $\mathscr{H}(t) < 0$. We model the headway distance from vehicle i to j as a $G/G/1$ queue, with each element of the queue size to be 1 m, and the instantaneous queue length represents the current headway distance between the two vehicles.

The evolvement of queue length is subject to the movement of vehicles i and j. Let v_i and v_j denote the mean velocity of vehicles i and j, respectively. Let a_i and a_j denote the variance of vehicle i and j's velocities, respectively. In other words, v_j and a_j represent the mean and variance of the (meter) arrival rate to the queue, and v_i and a_i represent the mean and variance of the (meter) service rate of the queue. By applying the $G/G/1$ model, we implicitly assume that the movement, i.e., the traveled distance of vehicles i and j, respectively, in the unit time is independent and identically distributed (i.i.d.), and is uncorrelated to that of each other. This is a working assumption in the multi-lane highway scenario with smooth and stable traffic. Notably, in the multi-lane highway with light traffic, vehicles i and j are free to adapt their velocities through accelarations/decelarations and lane changing which are subject to individual's driving habit [6]. In the case of the heavy traffic, as reported in [7], the headway distance between vehicles follows the Gaussian distribution which is related to the safety distance of different drivers. Moreover, for the ease of analysis, in this chapter, we consider the simple scenario in which the mobility of vehicles is stable in that the mean and variance of velocities are not changing over time. The assumption is practical as the variations of vehicle velocity on the highway are in a much larger time scale than the data transmissions.

We resort to the diffusion approximation [8] to evaluate the transient queue length, i.e., instantaneous headway distance $\mathscr{H}(t)$ between the two vehicles. The headway distance $\mathscr{H}(t)$ is modeled as a one-dimensional Wiener process (or Brownian motion) with the drift $\mu = v_j - v_i$ and variance $\sigma = a_j + a_j$. As such, within the infinitesimal interval Δt, the increment of $\mathscr{H}$ is normally distributed as

$$\Delta \mathscr{H}(t) = \mathscr{H}(t + \Delta t) - \mathscr{H}(t) = \mu \Delta t + \Theta \sqrt{\sigma \Delta t} \tag{3.1}$$

with Θ denoting the random variable following the unit normal distribution.

Let r denote the initial headway distance from vehicle i to vehicle j upon the time instant when the headway distance estimation, i.e., $\mathscr{H}(0) = r$ is initiated. Let $f_H(x;r,t)$ denote the probability density function (pdf) of $\mathscr{H}(t)$ at time t, conditional on the initial queue length, and

$$f_H(x;r,t) = \Pr\{x \leq \mathscr{H}(t) \leq x + \Delta x | \mathscr{H}(0) = r\}.$$

With the model in (3.1), $f_H(x,r,t)$ can be characterized by the Kolmogorov equation (alternatively known as Fokker–Planck equation) as

$$\frac{1}{2}\sigma\frac{\partial^2}{\partial x^2}f_H(x;r,t) + \mu\frac{\partial}{\partial x}f_H(x;r,t) = \frac{\partial}{\partial t}f_H(x;r,t) \tag{3.2}$$

subject to the initial condition of the headway distance,

$$f_H(x;r,0) = \delta(r), \tag{3.3}$$

with $\delta(\cdot)$ denoting the Dirac delta function.

Solving (3.2), we have [8]

$$f_H(x;r,t) = \frac{1}{\sqrt{2\pi\sigma t}}\exp\left\{-\frac{(x-r-\mu t)^2}{2\sigma t}\right\}. \tag{3.4}$$

Equation (3.4) represents the distribution of $\mathscr{H}(t)$ over time t provided the initial headway distance r and the statistics of vehicle velocity, and $f_H(x;r,t)$ follows the normal distribution with the mean $r + \mu t$ and variance σt.

3.2.2 Model of Wireless Channel Fading

Given the headway distance between the source and destination vehicles, we are ready to evaluate the physical layer transmission rate between them.

The vehicular communications operate on the exclusive DSRC frequency band between 5.850 and 5.925 GHz as mandated by US FCC. The 75 MHz DSRC spectrum is divided into seven 10 MHz channels with the rest 5 MHz reserved for future use. Of the seven channels, four channels can be used for the IP-based infortainment applications, with two channels intended for medium-range communication with the maximum transmit power to be 33 dBm, and another two channels for short-range communication with the maximum transmit power to be 23 dBm (for private vehicles) [9]. Over each channel, the physical layer operation of the vehicular communication is specified by the IEEE 802.11p standard, which adopts OFDM (Orthogonal Frequency Division Multiplexing) and works in the same manner of IEEE 802.11a physical layer. Four different modulation schemes (BPSK, QPSK, 16-QAM and 64-QAM) and three different FEC coding rates ($1/2, 2/3$ and $3/4$) are enabled, which leads to eight different transmission rates from 3 Mbps to 27 Mbps as shown in Table 3.1.

In this chapter, we adopt the measurement results in [10] to evaluate the physical layer capacity of V2V communications. In specific, let d denote the distance between vehicle i and j at time t, and $d = |\mathscr{H}(t)|$. The received signal strength as a function of d can be represented by the dual-slope piecewise-linear model [10], as

$$P(d) = \begin{cases} P(d_0) - 10\alpha_1 \log_{10}\left(\dfrac{d}{d_0}\right) + \chi_{\sigma_1}, & d_0 \leq d \leq d_c \\[2mm] P(d_0) - 10\alpha_1 \log_{10}\left(\dfrac{d_c}{d_0}\right) - 10\alpha_2 \log_{10}\left(\dfrac{d}{d_c}\right) + \chi_{\sigma_2}, & d > d_c \end{cases} \tag{3.5}$$

where $P(d_0)$ is the known signal strength at the reference distance d_0. α_1 and α_2 are the path loss exponents. d_c is the critical distance which can be evaluated as $d_c = \frac{4a_i a_j}{\lambda}$ with a_i, a_j denoting the antenna heights of vehicles i and j, respectively, and λ denoting the wavelength of the electromagnetic wave at 5.9 GHz. χ_{σ_1} and χ_{σ_2} zero-mean normally distributed random variables with the standard deviations σ_1 and σ_2, respectively.

Equation (3.5) characterizes the path loss and shadowing effects of the channel. Due to the typically short-range connections of V2V communications, a more accurate representation of the channel is to identify the fast fading. As reported in [10], the fast fading tends to be Rician at a short distance, and become severe as distance increases due to the gradual loss of the line-of-sight communication. Let s be a random variable, denoting the received signal envelope in the fast fading at vehicle j. The pdf of s can be represented by the Nakagami (m, Ω) distribution as

$$f_s(x; m, \Omega) = \frac{2m^m}{m^m \Gamma(m)} r^{2m-1} \exp\left(-\frac{m}{\Omega}x^2\right) \tag{3.6}$$

where Ω is the average received power before the envelop detection which can be evaluated as $\Omega = E[P(d)]$ with $P(d)$ shown in (3.5). m is the fading parameter. For $m = 1$, (3.6) reduces to Rayleigh distribution, and for $m = \frac{(K+1)^2}{2K+1}$, (3.6) becomes the Ricean fading with parameter K. $\Gamma(x) = \int_0^\infty e^{-t} t^{x-1} dt$ is the Gamma function. With (3.6), the cumulative density function (cdf) of the received power s^2 at vehicle j is

$$\Pr\{s^2 \leq x\} = 1 - \frac{\Gamma\left(m, \frac{m}{\Omega}x\right)}{\Gamma(\mu)} \tag{3.7}$$

where $\Gamma\left(m, \frac{m}{\varpi}x\right) = \int_{\frac{m}{\varpi}x}^\infty y^{\mu-1} \exp(-y) dy$.

With (3.7), the CDF of the signal to noise ratio (SNR) at the receiver is

$$\Pr\left\{\frac{s^2}{\varphi} \leq x\right\} = 1 - \frac{\Gamma\left(m, \frac{m}{\Omega}\varphi x\right)}{\Gamma(m)}, \tag{3.8}$$

where φ denotes the thermal noise power at the receiver.

We adopt the model in [11] to evaluate the channel modulation. In specific, we assume that the on-board wireless transceiver supports κ discrete modulation rates, denoted as $\Theta = \{c_1, c_2, \cdots, c_\kappa\}$ with $c_1 < c_2 < \cdots < c_\kappa$, and the transmission rates are adapted in the SNR-triggered manner. Specifically, each modulation rate, say c_k, is associated with a predefined threshold, say ϑ_k, where $\vartheta_{\kappa+1}$ is set ∞. The transmission rate c_k is adopted in the physical layer if the current SNR is above ϑ_k and smaller than ϑ_{k+1}. As such, based on (3.8), provided the distance of communication, the transmission rate c_k is selected with the probability

$$\Pr\{\mathscr{C} = c_k\} = \Pr\left\{\vartheta_k < \frac{s^2}{\varphi} \leq \vartheta_{k+1}\right\} \quad \text{for } 1 \leq k \leq \kappa - 1$$

$$= \frac{1}{\Gamma(\mu)}\left(\Gamma\left(m, \frac{m}{\Omega}\varphi\vartheta_k\right) - \Gamma\left(m, \frac{m}{\Omega}\varphi\vartheta_{k+1}\right)\right),$$

$$\Pr\{\mathscr{C} = c_k\} = \frac{\Gamma\left(m, \frac{m}{\omega}\varphi\vartheta_\kappa\right)}{\Gamma(m)} \qquad \text{for } k = \kappa, \tag{3.9}$$

and with the rest probability

$$\Pr\{\mathscr{C} = 0\} = 1 - \sum_{k=1}^{\kappa} \Pr\{\mathscr{C} = c_k\}.$$

3.2.3 Channel Contention Model

The vehicular networks adopt the contention-based MAC to resolve the channel contentions among parallel transmissions. In this part, we evaluate the MAC throughput $\mathscr{R}$ from i to j.

Let $\mathscr{N}$ denote the number of vehicles, including vehicle i, which are contending the channel for transmissions in the highway section. We assume that $\mathscr{N}$ follows the Poisson distribution as

$$f_n(\mathscr{N}) = \frac{(\rho S)^{\mathscr{N}}}{\mathscr{N}!}\exp(-\rho S) \tag{3.10}$$

where ρ is the traffic density, defined as vehicles/m. S denotes the carrier sensing range of vehicles. The Poisson distribution of vehicles on the highway has been reported in [12] based on the analysis of real-world trace. This model is also widely used in literature, e.g., [13].

We assume that the IEEE 802.11b DCF (distributed coordination function) scheme is applied for the MAC scheduling with the RTS/CTS scheme adopted to eliminate the hidden terminals of transmissions. Let W denote the minimum contention window size used in the exponential backoff of vehicle i. Let τ denote the average transmission probability of each vehicle, and

$$\tau = \frac{1}{W/2 + 1}. \tag{3.11}$$

The probability of the successful transmission from vehicle i to vehicle j, given that the vehicle i transmits, is

$$P_{\mathrm{suc}} = (1 - \tau)^{\mathcal{N}-1}. \tag{3.12}$$

The MAC throughput from vehicle i to vehicle j is therefore

$$\mathcal{R} = \tau P_{\mathrm{suc}} \frac{\mathrm{FL}_i}{T} \tag{3.13}$$

where FL_i is the frame/packet length of vehicle i, including the payload and the packet header. T is the average length of a time slot in DCF, mathematically

$$T = (1 - P_{\mathrm{tran}})\,\mathrm{SlotTime} + (P_{\mathrm{tran}} - P_{\mathrm{suc}})\,T_{\mathrm{cld}} + P_{\mathrm{suc}} T_{\mathrm{suc}},$$

where P_{tran} is the probability that the channel is busy for transmission and $P_{\mathrm{tran}} = 1 - (1 - \tau)^n$. P_{suc} is the probability of successful transmission when channel is busy, and $P_{\mathrm{suc}} = n\tau(1 - \tau)^{n-1}$. T_{cld} and T_{suc} are the average time of collided and successful transmissions, respectively. Mathematically, we have

$$\begin{cases} T_{\mathrm{cld}} = \mathrm{RTS} + \mathrm{DIFS} + \mathrm{SlotTime}, \\ T_{\mathrm{suc}} = \mathrm{RTS} + 3 \times \mathrm{SIFS} + 4 \times \mathrm{SlotTime} + \mathrm{CTS} \\ \qquad + \dfrac{E(\mathrm{FL})}{E(C)} + \mathrm{ACK} + \mathrm{DIFS}, \end{cases} \tag{3.14}$$

where SlotTime is the unit slot time of DCF backoff. SIFS and DIFS are predefined time intervals reserved for the DCF signallings and operations. RTS, CTS and ACK represent the time interval of RTS, CTS and ACK transmissions, respectively. Here, we assume that the RTS/CTS scheme is adopted with DCF to eliminate the hidden terminals during the transmission.

$E(\mathrm{FL})$ in (3.14) is the average frame length, including the payload and packet header, of transmissions. Given that $\mathcal{N}$ vehicles are contending the channel, we have

$$E(\mathrm{FL}) = \frac{1}{\mathcal{N}}\mathrm{FL}_i + \frac{\mathcal{N}-1}{\mathcal{N}}\widehat{\mathrm{FL}}, \tag{3.15}$$

with $\widehat{FL}$ being the average frame length of the other vehicles. In this chapter, we assume that the frame length of different vehicles follow the same and known distribution.

$E(C)$ in (3.14) is the average physical layer transmission rate of vehicles. Given the transmission rate of vehicle i to be $\mathscr{C}$, we have

$$E(C) = \frac{1}{\mathscr{N}}\mathscr{C} + \frac{\mathscr{N}-1}{\mathscr{N}}\hat{C}, \tag{3.16}$$

where $\hat{C}$ is the average transmission rate of the other vehicles. We assume that transmission distance among each source destination pair in the network is uniformly distributed[2] within the carrier sensing range S, and accordingly

$$\hat{C} = \frac{1}{S}\int_0^S c_k \Pr\{\mathscr{C} = c_k\} dx. \tag{3.17}$$

3.2.4 Data Download Evaluation

We are now ready to evaluate the data amount $\mathscr{A}$ that can be transmitted from vehicle i to vehicle j. Within a time period of Υ ($\Upsilon \geq 0$), the integrated data volume transmitted from i to j is evaluated by integrating the MAC throughput $\mathscr{R}$ over the range $[0, \Upsilon]$ as

$$\mathscr{A} = \int_0^\Upsilon \mathscr{R} dt. \tag{3.18}$$

Note that (3.18) involves the integration over a random process. To simplify the analysis, we only derive the mean and an upper bound of the variance of $\mathscr{A}$ as

$$\mathbb{E}(\mathscr{A}) = \int_0^\Upsilon \mathbb{E}(\mathscr{R}) dt \tag{3.19}$$

where

$$\mathbb{E}(\mathscr{R}) = \int_{-\infty}^\infty \sum_{n=1}^\infty \sum_{k=1}^K P(\mathscr{C} = c_k) \mathscr{R} f_n(\mathscr{N}) f_H(x; r, t) dx,$$

and

$$\mathbb{V}(\mathscr{A}) \leq \Upsilon \int_0^\Upsilon \mathbb{E}(\mathscr{R}^2) dt - \mathbb{E}(\mathscr{A})^2, \tag{3.20}$$

---

[2]This also implies that the content transmissions among vehicles are independent of their distance.

where

$$\mathbb{E}\left(\mathscr{R}^2\right) = \int_{-\infty}^{\infty} \sum_{n=1}^{\infty} \sum_{k=1}^{K} P\left(\mathscr{C} = c_k\right) \mathscr{R}^2 f_n\left(\mathscr{N}\right) f_H\left(x; r, t\right) dx.$$

The derivations of (3.19) and (3.20) are shown in the Appendix.

3.3   Call Admission Control

In this section, we devise an admission control scheme at source vehicles to prevent content transmissions which are unlikely to be finished within the short-lived connection time.

Let F_{ij} denote the file size to be transmitted from vehicle i to vehicle j. Before the file transmission, we assume that a three-tuple $\{\text{Location}_j, v_j, a_j\}$ is notified to vehicle i by vehicle j, piggybacked by the download request of vehicle j, where Location_j is the GPS location of vehicle j. As such, before the mass file transmission, vehicle i first examines the following inequality as

$$\Pr\left\{\mathscr{A} > F_{ij}\right\} > \xi, \tag{3.21}$$

where ξ is predefined and $0 << \xi < 1$. If (3.21) can be satisfied, which indicates that the content transmission can be completed with high probability, the download request of vehicle j will be approved with ensuing data transmissions from i to j. Otherwise, the download request will be rejected so as to prevent the potential transmission of fragment contents and avoid waste of bandwidth.

As the distribution of $\mathscr{A}$ is unknown, we apply the Chebyshev inequality to relax (3.21). According to the one-sided Chebyshev inequality, we have

$$\Pr\left\{\mathscr{A} \leq F_{ij}\right\} \leq \frac{\mathbb{V}\left(\mathscr{A}\right)}{\mathbb{V}\left(\mathscr{A}\right) + \left[\mathbb{E}\left(\mathscr{A}\right) - F_{ij}\right]^2} \tag{3.22}$$

with $F_{ij} < \mathbb{E}\left(\mathscr{A}\right)$.

By substituting (3.20) and (3.22) into (3.21), we have that (3.21) can be satisfied if the following inequality is satisfied

$$\frac{\Upsilon \int_0^\Upsilon \mathbb{E}\left(\mathscr{R}^2\right) dt - \mathbb{E}\left(\mathscr{A}\right)^2}{\Upsilon \int_0^\Upsilon \mathbb{E}\left(\mathscr{R}^2\right) dt - \mathbb{E}\left(\mathscr{A}\right)^2 + \left[\mathbb{E}\left(\mathscr{A}\right) - F_{ij}\right]^2} \leq 1 - \xi, \tag{3.23}$$

where Υ is a predefined scalable, representing the deadline of content transmission required by the applications at the receiver j.

By implementing (3.23) in the admission control, the request of content transmission is approved by vehicle i if both of the following conditions are met:

(1) $F_{ij} < \mathbb{E}(\mathscr{A})$, and (2) (3.23) is satisfied. Note that as (3.22) overestimates $\Pr\{\mathscr{A} \le F_{ij}\}$, the resultant admission control scheme is more conservative than that adopts (3.21). Moreover, using an upper bound of $\mathbb{V}(\mathscr{A})$ as represented by (3.20) in (3.23) will also make the admission control scheme conservative.

3.4 Simulation Verification

In this section, we examine the accuracy of the analytical model and effectiveness of the described admission control scheme using simulations.

3.4.1 Simulation Setup

Our simulations are based on a customer simulator coded in C++, which is adapted from [14, 15]. In each simulation run, we simulate file transmissions among vehicles on a linear highway topology with the following configurations:

Vehicle Mobility: We simulate 1,000 vehicles in a highway road section. At the commencement of each simulation run, vehicles are placed on the road following Poisson distribution; the headway distance between neighboring vehicles follows the exponential distribution with mean value of 60 m. The initial density of vehicles is thus $\rho = 1/60$ car/m. When the simulation proceeds, vehicles adapt their velocity at each unit time following Normal distribution with the mean value uniformly distributed within $[70, 130]$ km/h and the standard deviation uniformly distributed within $[21, 39]$ km/h. A similar configuration of vehicle mobility is also used in [6].

File Transmission: Along its trajectory, each vehicle iteratively requests data files to download from the neighboring vehicles. In specific, upon each iteration of file transmission, a vehicle (as receiver) first randomly selects a source node from its one-hop neighbors within the communication range of 300 m. The selected source node then generates a content file with the size uniformly distributed within the range $[1, 10]$ MB and transmit to the receiver vehicle over the DSRC radio. The file is transmitted in a packet flow with the frame length FL following the distribution as $P(\text{FL}) = 0.2 \times \delta(\text{FL} - 200) + 0.25 \times \delta(\text{FL} - 800) + 0.35 \times \delta(\text{FL} - 1,200) + 0.2 \times \delta(\text{FL} - 1,500)$. The file transmission terminates if either one of the following three events occurs: (1) the entire file has been transmitted successfully to the receiver; (2) the receiver vehicle moves outside the communication range (i.e., 300 m) of the source vehicle; (3) a deadline of content transmission $\Upsilon = 100$ s is reached. The termination of a file transmission then initiates a new iteration of file transmission following the same procedure as above.

Channel Fading: Each vehicle is equipped with a single transceiver operating on the DSRC radio. Based on the value of received SNR, the transmitter adapts the modulation scheme and physical layer transmission rate according to SNR thresholds shown in Table 3.1 [16]. The thermal noise power is set to be -96 dBm

Table 3.1 DSRC data rate and SNR threshold

SNR threshold (dB)	5	6	8	11	15	20	25	N/A
Data rate (Mbps)	3	4.5	6	9	12	18	24	27

Table 3.2 Fading parameter m over distance d (in meter)

d	≤ 5.5	≤ 13.9	≤ 35.5	≤ 90.5	≤ 230.7	≤ 588
m	4.07	2.44	3.08	1.52	0.74	0.84

which is same as [14]. The received signal is deteriorated by the Nakagami-m fading as (3.6). In (3.6), the fading parameter m is a function of the transmission distance is based on the measurement in [10] as shown in Table 3.2. The average received power is determined by (3.5) in which $\alpha_1 = 2.1$, $\alpha_2 = 3.8$ based on [10] and the received power at reference distance $d_0 = 100$ m is evaluated by the two-ray ground reflection model as $P(d_0) = P_t G_t G_r \frac{h_t^2 h_r^2}{d_0^4 L}$ with the transmission power $P_t = 23$ dBm (short-range), gain of the transmitter (receiver) antenna $G_t = 1$ ($G_r = 1$), height of the transmitter (receiver) antenna $h_t = 1$ m ($h_r = 1$ m) and system loss factor $L = 1$.

MAC Contention: The channel access of packet transmissions is scheduled by the DCF MAC with RTS/CTS handshake. The minimum contention window of vehicles is set to be $W = 32$, and the carrier sensing range S is sent to be 500 m. Other MAC layer parameters are as follows: SlotTime $= 13\,\mu$s, SIFS $= 32\,\mu$s, DIFS $= 32\,\mu$s, RTS transmission time $= 53\,\mu$s, CTS transmission time $= 37\,\mu$s and ACK transmission time $= 37\,\mu$s.

3.4.2 Distribution of Headway Distance

In the first experiment, we validate the accuracy of (3.4) on predicting of the distribution of headway distance after time period t. To achieve this goal, we insert two vehicles i and j (assuming i transmits to j) into the network with the controllable initial headway distance r and velocities. Unless otherwise mentioned, we set $v_i = 97.2$ km/h, $a_i = 30.6^2$ and $v_j = 90$ km/h, $a_j = 27^2$ in this experiment. The averaged results over 300 simulation runs are reported and compared with the analysis derived from (3.4).

Figure 3.3 shows the CDF of the headway distance $\mathcal{H}(t)$ when $t = 100$ s with the different values of initial distance r. By increasing r, it can be seen that the curves shift to the right hand side, indicating the increasing mean value of $\mathcal{H}(t)$. The variance of $\mathcal{H}(t)$ is not related to r as indicated by (3.4) and simulations. Figure 3.4 shows the CDF of $\mathcal{H}(t)$ at different time t with $r = -100$ m (i.e., vehicle i is ahead of vehicle j in the moving direction). It can be seen that by increasing t, both the mean and variance of $\mathcal{H}(t)$ increase. With the communication range of vehicles typically to be 300 m, from Figure 3.4, it can be seen that at $t = 90$ s, the probability that vehicles i and j are connected is 0.7 and this probability reduces to almost 0

Fig. 3.3 CDF of $\mathcal{H}_{ij}(t)$ at $t = 100$ s with different initial headway distance r

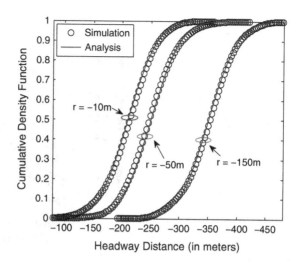

Fig. 3.4 CDF of $\mathcal{H}_{ij}(t)$ with initial headway distance $r = -100$ m at different time t

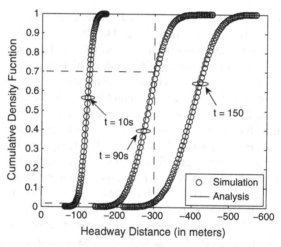

when $t = 150$ s. Figure 3.5 shows the CDF of $\mathcal{H}(t)$ with different a_j (variance of vehicle j's velocity). By increasing a_j, vehicle j adapts its velocity more intensively over time. This accordingly leads to the increased variance of $\mathcal{H}(t)$ and reduced probability of connection between i and j with fixed r as indicated by Figure 3.5.

3.4.3 Download Performance During Connection Time

In the second experiment, we investigate on the accuracy of (3.19) and (3.20) on evaluating the mean and variance of the transmitted data volume among vehicles. To this end, we investigate on the same pair of vehicles i and j in the previous experiment and let vehicle i transmit to vehicle j with the packet length to be 1,000

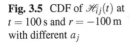

Fig. 3.5 CDF of $\mathscr{H}_{ij}(t)$ at $t = 100\,$s and $r = -100\,$m with different a_j

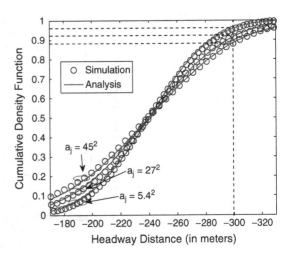

bytes. We report the averaged results over 300 simulation runs and compare the analysis derived from (3.19) and (3.20).

Figure 3.6 shows the mean data volume that can be transmitted from vehicle i to vehicle j as a function of time when the initial headway distance from vehicle i to vehicle j is $r = 50\,$m. The data volume at each instant represents the data transmitted from time 0 until the corresponding time instant. As we can see from Figure 3.6, with the time increasing, the mean volume of transmitted data increases as download time increases whereas the rate of increment reduces. This is because that as time elapses, the distance between i to j increases and the transmission rate of vehicle i reduces accordingly. Figure 3.6 shows the variance of the data volume that can be transmitted from vehicle i to vehicle j when $r = 50\,$m. Note that (3.20) represents an upper bound of $\mathbb{V}(\mathscr{A})$. As we can see from Figure 3.7, with time increasing, the gap of the bound increases. This is because that the error of estimation accumulates over time. Figure 3.8 shows the mean data volume transmitted from i to j with different values of initial distance r. As we can see, when $r = 150\,$m, more data can be transmitted compared with that with $r = 10\,$m. This is because that as $v_i > v_j$ the high-rate connection between i and j is longer in the case when $r = 150\,$m than that when $r = 10\,$m.

3.4.4 Validation of Call Admission Control

In the last experiment, we examine the effectiveness of the described admission control scheme. We investigate on a randomly selected vehicle node in the network and set the mean and standard deviation of its velocity to be 97.2 km/h and 30.6 km/h, respectively, unless otherwise mentioned. The selected vehicle continuously subscribes to download files from neighboring vehicles in the communication

Fig. 3.6 Mean data volume transmitted from i to j until the simulation time

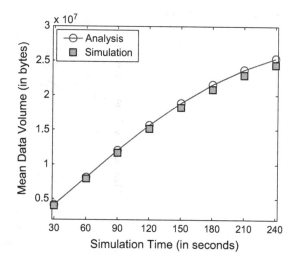

Fig. 3.7 Variance of the data volume transmitted from i to j until the simulation time

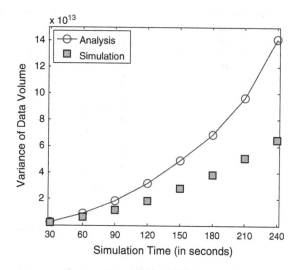

range following the file transmission behavior as aforementioned. The deadline of content transmission Υ in (3.23) is set to be 100 s and ξ is set to be 0.9. In what follows, we show the download performance of the selected vehicle with and without the admission control mechanism applied, respectively.

Figure 3.9 shows the fraction of fragment contents in the overall data volume transmitted when the admission control mechanism is not applied, with different standard deviations of the selected vehicle's velocity. As we can see in the simulated scenario, around 23–37 % of the overall transmitted data are wasted for the delivery of fragment contents. By increasing the standard deviation of the selected vehicle's velocity, more fragment content transmissions are observed as in this case the mobility of the selected vehicle changes more intensively, leading to the more

Fig. 3.8 Variance of the data volume transmitted from i to j until the simulation time with different r at $t = 0$

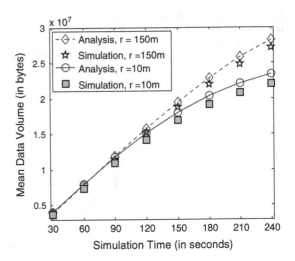

Fig. 3.9 Fraction of fragment contents in the overall data volume transmitted with different standard deviation of the selected vehicle

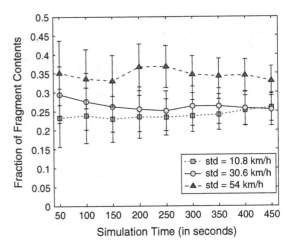

dynamic and unreliable V2V connections. Figure 3.10 shows the overall data received by the selected vehicle, together with the valid data of entire content downloads, when the admission control scheme is not applied. Similar to Figure 3.9, by increasing the standard deviation of the selected vehicle's velocity, we observe the decreased data volume received by the vehicle. This also attributes to the increase of channel dynamics due to the enhanced mobility.

Figure 3.11 shows the fraction of fragment contents in the overall transmissions during the intervals of [0, 100 s] and [0, 450 s], respectively, with and without admis-

Fig. 3.10 Valid and overall
data volume transmitted

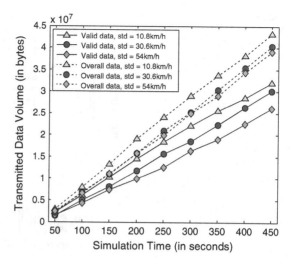

Fig. 3.11 Fraction of
fragment contents with and
without admission control
(CAC) mechanism applied

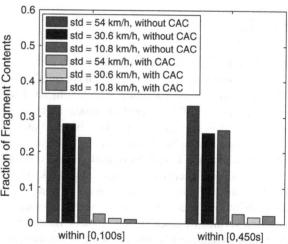

sion control applied. As we can see, with the admission control mechanism applied,
the fraction of fragment contents downloaded is below 3% of the overall data
received. With the standard deviation of the selected vehicle's velocity reducing,
the fraction of fragment contents downloaded reduces. Figure 3.12 shows the valid
data volume downloaded by the selected vehicle within the intervals of [0,100 s]
and [0,450 s], respectively, with and without admission control applied. With the
transmission of fragment contents dramatically reduced due to the admission
control, the valid data receives increases around 50–60% of the case without
admission control.

Fig. 3.12 Valid received data volume with and without admission control (CAC) mechanism applied

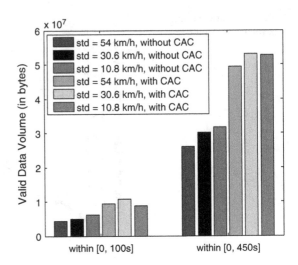

3.5 Summary

We conclude the chapter by emphasising that at the foundation of efficient vehicular communications lies the basic requirement of effective and efficient transmissions of *intact content files* to the highly mobile vehicles. Due to the dynamic nature of the volatile and spotty V2V channels, to guarantee the integrity of inter-vehicle content transmissions is very challenging but foremost and key to enable efficient media applications to vehicles. In this chapter, we have described a theoretical treatment on provisioning the integrity-oriented content transmission. To this end, we have presented a comprehensive analytical framework to evaluate the performance of data content transmissions over the dynamic V2V channel. The presented model captures the node mobility, channel fading and MAC contentions in one framework, and has been verified by extensive simulations. Based on the presented model, we have described an admission control scheme at the transmitter to filter the transmission requests which are unlikely to be finished during the transient connection time. Using simulations, we have shown that the presented scheme can help save around 30 % of the network bandwidth in the simulated scenario.

To extend this work, there may exist two interesting directions. Firstly, this chapter only considers the content transmission in the single-hop V2V connections. When multi-hop transmissions are involved, the problem would be more generic yet challenging and deserves further study. Secondly, this chapter does not consider the multi-party download case in which a vehicle is possible to retrieve the same contents from multiple upload nodes. It is also interesting to extend the presented work by considering the content distribution applications with multi-party download available.

3.6 Appendix

According to (3.18), we have

$$\mathbb{E}(\mathscr{A}) = \mathbb{E}\left(\int_0^{\Upsilon} \mathscr{R} dt\right) = \int_0^{\Upsilon} \mathbb{E}(\mathscr{R}) dt. \tag{3.24}$$

$\mathbb{E}(\mathscr{R})$ can be derived based on the conditional expectation as

$$\mathbb{E}(\mathscr{R}) = \mathbb{E}_d \mathbb{E}_n \mathbb{E}_c (\mathscr{R}; \mathscr{N}, d) \tag{3.25}$$

where $\mathbb{E}_c(\cdot)$ is the expectation on the physical layer transmission rate. $\mathbb{E}_n(\cdot)$ is the expectation on the number of vehicles in the carrier sensing range. $\mathbb{E}_d(\cdot)$ is the expectation on the distance between vehicles i and j. Note that the distribution of $\mathscr{N}$ and d are independent.

Given (3.9), we have

$$\mathbb{E}_c(\mathscr{R}; \mathscr{N}, d) = \sum_{k=1}^{\kappa} P\{\mathscr{C} = c_k\} \mathscr{R}. \tag{3.26}$$

Given the Poisson distribution of $\mathscr{N}$ in (3.10)

$$\mathbb{E}_n \mathbb{E}_c(\mathscr{R}; \mathscr{N}, d) = \sum_{\mathscr{N}=1}^{\infty} \mathbb{E}_c(\mathscr{R}; \mathscr{N}, d) f_n(\mathscr{N}). \tag{3.27}$$

Given the distribution of the headway distance specified in (3.4), we have

$$\mathbb{E}_d \mathbb{E}_n \mathbb{E}_c(\mathscr{R}; \mathscr{N}, d) = \int_{-\infty}^{\infty} f_H(x; r, t) \mathbb{E}_n \mathbb{E}_c(\mathscr{R}; \mathscr{N}, |x|) dx. \tag{3.28}$$

Substituting (3.25)–(3.28) into (3.24), we have (3.19).

To derive (3.20), we have

$$V(\mathscr{A}) = \mathbb{E}(\mathscr{A}^2) - \mathbb{E}(\mathscr{A})^2. \tag{3.29}$$

where

$$\mathbb{E}(\mathscr{A}^2) = \mathbb{E}\left(\int_0^{\Upsilon} \mathscr{R}_t dt \int_0^{\Upsilon} \mathscr{R}_{\tau} d\tau\right) \tag{3.30}$$

with $\mathscr{R}_t$ and $\mathscr{R}_{\tau}$ following the same distribution as $\mathscr{R}$ at time t and τ, respectively.

To evaluate the integral in (3.30), we partition the time duration $[0, \Upsilon]$ into χ slots with each slot $\Delta t = \frac{\Upsilon}{\chi}$. According to the definition of Riemann integral, we have

$$\int_0^\Upsilon \mathscr{R}_t dt = \lim_{\chi \to \infty} \sum_{i=1}^\chi \mathscr{R}_{i\Delta t} \Delta t \tag{3.31}$$

In the similar manner, we have

$$\int_0^\Upsilon \mathscr{R}_\tau d\tau = \lim_{\chi' \to \infty} \sum_{j=1}^{\chi'} \mathscr{R}_{j\Delta \tau} \Delta \tau \tag{3.32}$$

with $\Delta \tau = \frac{\Upsilon}{\chi'}$.

Substituting (3.31) and (3.32) into (3.30), we have

$$\mathbb{E}\left(\mathscr{A}^2\right) = \lim_{\chi \to \infty} \lim_{\chi' \to \infty} \mathbb{E}\left(\sum_{i=1}^\chi \sum_{j=1}^{\chi'} \mathscr{R}_{i\Delta t} \mathscr{R}_{j\Delta \tau} \Delta t \Delta \tau\right) \tag{3.33}$$

$$= \lim_{\chi \to \infty} \lim_{\chi' \to \infty} \left(\sum_{i=1}^\chi \sum_{j=1}^{\chi'} \mathbb{E}\left(\mathscr{R}_{i\Delta t} \mathscr{R}_{j\Delta \tau}\right) \Delta t \Delta \tau\right)$$

Since $\mathscr{R}_{i\Delta t}$ and $\mathscr{R}_{j\Delta t}$ are positive real numbers, we have

$$\mathscr{R}_{i\Delta t} \mathscr{R}_{j\Delta \tau} \leq \frac{1}{2}\left(\mathscr{R}_{i\Delta t}^2 + \mathscr{R}_{j\Delta \tau}^2\right).$$

Substituting it into (3.33), we have

$$\mathbb{E}\left(\mathscr{A}^2\right) \leq \lim_{\chi \to \infty} \lim_{\chi' \to \infty} \frac{1}{2} \sum_{i=1}^\chi \sum_{j=1}^{\chi'} \mathbb{E}\left(\mathscr{R}_{i\Delta t}^2 + \mathscr{R}_{j\Delta \tau}^2\right) \Delta t \Delta \tau$$

$$= \lim_{\chi \to \infty} \lim_{\chi' \to \infty} \sum_{i=1}^\chi \sum_{j=1}^{\chi'} \mathbb{E}\left(\mathscr{R}_{i\Delta t}^2\right) \Delta t \Delta \tau \tag{3.34}$$

$$= \Upsilon \int_0^\Upsilon \mathbb{E}\left(\mathscr{R}^2\right) dt$$

where

$$\mathbb{E}\left(\mathscr{R}^2\right) = \int_{-\infty}^\infty \sum_{n=1}^\infty \sum_{k=1}^K P\{\mathscr{C} = c_k\} \mathscr{R}^2 f_n\left(\mathscr{N}\right) f_H\left(x; r, t\right) dx$$

Substituting (3.34) into (3.29), we have (3.20).

References

1. T. Zahn, G. O'Shea, and A. Rowstron, "Feasibility of Content Dissemination Between Devices in Moving Vehicles," in *Proc. of ACM CoNEXT*, 2009.
2. M. Boban, T. T. V. Vinhoza, M. Ferreira, J. Barros, and O. K. Tonguz, "Impact of Vehicles as Obstacles in Vehicular Ad Hoc Networks," *IEEE Journal on Selected Areas in Communications*, vol. 29, pp. 15–28, Jan. 2011.
3. B. Yu and F. Bai, "ETP: Encounter Transfer Protocol for Opportunistic Vehicle Communication," in *Proc. of IEEE Infocom*, 2011.
4. A. D. May, *Traffic Flow Fundamentals*. Prentice-Hall, 1990.
5. T. H. Luan, X. Shen, and F. Bai, "Integrity-Oriented Content Transmission in Highway Vehicular Ad Hoc Networks," in *Proc. of IEEE Infocom*, 2013.
6. S. Yousefi, E. Altman, R. El-Azouzi, and M. Fathy, "Analytical Model for Connectivity in Vehicular Ad Hoc Networks," *IEEE Transactions on Vehicular Technology*, vol. 57, pp. 3341–3356, Nov. 2008.
7. S. Ahn, M. J. Cassidy, and J. Laval, "Verification of a Simplified Car-Following Theory," *Transportation Research Part B: Methodological*, vol. 38, no. 5, pp. 431–440, 2004.
8. D. R. Cox and H. D. Miller, *The Theory of Stochastic Processes*. Chapman & Hall/CRC, 1977.
9. H. Hartenstein and K. Laberteaux, *VANET: Vehicular Applications and Inter-Networking Technologies*. Wiley, 2010.
10. L. Cheng, B. E. Henty, D. D. Stancil, F. Bai, and P. Mudalige, "Mobile Vehicle-to-Vehicle Narrow-band Channel Measurement and Characterization of the 5.9 GHz Dedicated Short Range Communication (DSRC) Frequency Band," *IEEE Journal on Selected Areas in Communications*, vol. 25, pp. 1501–1516, Oct. 2007.
11. J. Yoo, B. S. C. Choi, and M. Gerla, "An Opportunistic Relay Protocol for Vehicular Road-side Access with Fading Channels," in *Proc. of IEEE ICNP*, 2010.
12. F. Bai and B. Krishnamachari, "Spatio-Temporal Variations of Vehicle Traffic in VANETs: Facts and Implications," in *Proc. of ACM VANETs*, 2009.
13. W. Zhang, Y. Chen, Y. Yang, X. Wang, Y. Zhang, X. Hong, and G. Mao, "Multi-Hop Connectivity Probability in Infrastructure-Based Vehicular Networks," *IEEE Journal on Selected Areas in Communications*, vol. 30, pp. 740–747, May 2012.
14. J. J. Haas, Y. C. Hu, and K. P. Laberteaux, "Real-World VANET Security Protocol Performance," in *Proc. of IEEE Globecom*, 2009.
15. Q. Chen, F. Schmidt-Eisenlohr, D. Jiang, M. Torrent-Moreno, L. Delgrossi, and H. Hartenstein, "Overhaul of IEEE 802.11 Modeling and Simulation in NS-2," in *Proc. of ACM MSWiM*, 2007.
16. D. Jiang, Q. Chen, and L. Delgrossi, "Optimal Data Rate Selection for Vehicle Safety Communications," in *Proc. of ACM VANET*, 2008.

Chapter 4
Large-Volume Content Distribution in Vehicular Networks: Adaptive Playout from User's Perspective

This chapter considers the distribution large-volume contents, like video files, to vehicles. To reduce the start-up delay of the media presentation, real-time streaming is normally adopted which initiates the media playback during the download of content files. However, the opportunistic connectivity and dynamic download throughput of vehicles severely threats to streaming service quality perceived by users, and therefore how to accommodate the network dynamics and deliver the best perceived service quality to users are the focus of this chapter.

4.1 Introduction

The live distribution of large-volume video contents, such as live and on-demand video streaming, has achieved significant success and great popularity world-wide. For example, it is reported that there were 200,000 online users simultaneously watching the annual spring festival gala show on Chinese New Year Eve in 2006 using a peer-to-peer video streaming software [1]. As disclosed in Sandvine traffic report [2] in 2013, Netflix and YouTube dominate over half of downstream Internet traffic in North America. Notably, due to the constrained space and barely any entertainment available, it is highly expected that live media content distribution will be even more demanded by vehicular users. This thus calls for the efficient design of live media streaming systems over the vehicular networks.

The real-time media streaming services typically have stringent QoS requirements to maintain user's satisfaction. High quality media streaming over the spotty and highly dynamic vehicular communication links represents multiple fundamental challenges in engineering. For instance, the download links of vehicles are by nature dynamic with severe channel fading due to the high vehicle mobility and opportunistic connections to the roadside infrastructure. This leads to the dramatic changes of end-to-end throughput and download delays. As video packets have strict deadlines of presentation, the varying network delays may result in the missing of

T.H. Luan et al., *Enabling Content Distribution in Vehicular Ad Hoc Networks*,
SpringerBriefs in Computer Science, DOI 10.1007/978-1-4939-0691-8_4,
© The Author(s) 2014

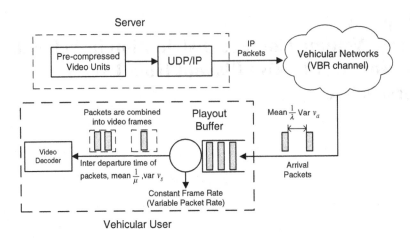

Fig. 4.1 Process of video streaming

playback deadline and consequently the jerkiness or even frozen of video playback. Therefore, to understand how the network dynamics, which is instinct to vehicular networks, affect the perceived video quality by on-board users and effectively accommodate the dynamic network delays in the video playback is crucial.

There are a variety of techniques appeared in the literature to address the problem of packet video delivery over dynamic and time-varying channels [3, 4], namely variable bit-rate (VBR) channels, including the rate-distortion optimized packet scheduling [5] and routing [6–8], power control and adaptive coding at the transmitter [9–11], playback rate and strategy adaption at the receiver [12, 13], etc. However, most of these approaches focus on the network-side issues which consider the network optimization in terms of throughput, delay and delay jitter. In this chapter, we describe the proposal to investigate from the end user's perspective.

We describe the proposal which analyzes and optimizes the user perceived video quality in terms of the start-up delay, fluency of video playback and packet loss rate, when downloading media flows through the VBR channel like vehicular networks. By intelligently using the user-side resources which are dependent on network dynamics, the described proposal designs adaptive quality-driven video streaming schemes towards the maximal user utility. To fold the journey, we start by describing an analytical framework to reveal the impacts of network dynamics on the video quality perceived by the end user. In specific, to overcome the network dynamics, a playout buffer is usually deployed at the receiver as shown in Figure 4.1. The playout buffer stores the received video packets and delays the initial media playout for a short period until a certain playback threshold is reached. This short period constitutes the start-up delay. During the media playout, the packets are discharged from the playout buffer and injected to the media player for playback. As long as the playout buffer is non-empty, the continuous media playout is always guaranteed. In this manner, the playout buffer can absorb the network dynamics by reserving some packets in the cache. For this sake, the playout buffer is also referred to as the dejitter buffer. Since the occupancy of the playout buffer is closely related to the user's viewing experience in terms of start-up delay and smoothness of playback, we study the video quality by analyzing the evolution of the queue length under the network dynamics. Secondly, we describe adaptive schemes to optimally determine the playback threshold driven by the users' required video quality. We strike a trade-off among the start-up delay, smoothness of playback and video packet loss, by adjusting the playback threshold. Specifically, to ensure smooth media presentation, enough packets should be cached in the playout buffer to absorb the variations of packet arrivals in the future. This, however, may incur an intolerable long waiting time to end users. Meanwhile, to prevent packet loss due to overflow of the finite buffer, the queue length should be maintained at a relatively low level with less packets buffered initially. However, low queue occupancy may cause frequent playback interruptions. Thus, the playback threshold should be adaptively and optimally determined under the specific requirements of video quality metrics. Based on the analytical framework, the described proposal formulates the playback threshold selection as a stochastic optimization problem driven by the specific video quality requirements, and provide the optimal solutions to guarantee the stochastic video performance to end users.

The remainder of this chapter is organized as follows. We first provide a brief review of the related literature, and then present the analytical framework in details by considering both cases when the size of playout buffer is infinite and finite, respectively. We then describe multiple playout buffer adaption schemes with different network performance. The described proposal is then validates by extensive simulations. Lastly, we close this chapter with a summary. Note that without any assumption on the underlying VBR channels, the described proposal can actually be used in a broad networks besides vehicular networks. We also describe a proposal which applies the results derived in this chapter to optimize the live media streaming in the V2R scenario.

4.2 Related Literature

Streaming media over unreliable and time-varying VBR channels has attracted an extensive research attention in the last decade. Various network-adaptive schemes have been proposed [3]. The proposal described in this chapter belongs to the scope of end-system centric solutions [4] which adapt the video from the receiver's perspective.

The end-system centric solutions refer to the adaptive video streaming mechanisms which adaptively modify the visual quality via the playback rate control or playout buffer management at the end-systems based on the occupancy of playout buffer or user's available bandwidth [12]. Liu et al. [13] propose an end-to-end playback rate adaptation scheme based on the layer coding technique. Each receiver actively measures its local available bandwidth and pass that to the server. Based on the echo information, the server then determines the appropriate number of layered streams conveyed to users and hence adapts the video compression rate according to the available bandwidth. By doing so, the visual quality degrades with enhanced compression ratio when the end-to-end bandwidth is insufficient; nevertheless, users can enjoy smooth playback. Galluccio et al. [14] describe an adaptive MPEG video streaming framework in which the wireless channel is modeled as a Rayleigh fading channel represented by a FSMC (finite state Markov chain). By analyzing the channel status via the Markovian model, the available bandwidth can be computed and the appropriate video playback rate is determined accordingly. Similar approach is adopted in [9] where channel coding is adapted in different channel conditions which are evaluated using FSMC.

To distribute the computation burden to the end users and hence enhance the network scalability, some approaches have been proposed to adapt the video playback at the end users. Kalman et al. [15] introduce the adaptive media playback (AMP) scheme at the end user, which can adaptively tune the video playout rate according to the playout buffer occupancy to ensure the smooth video playback. In specific, when the occupancy of playout buffer is above some threshold, the video playback rate will be increased to avoid the overflow of playout buffer. This leads to the effects of fast forward to the users. Laoutaris et al. [16] adopt the same mechanism but use the Markov decision process (MDP) to optimally determine the video playback rate at different channel conditions.

Another prevailing way of adaptive video streaming is by playout buffer management. In this scenario, the key issue is how to optimally determine the playback threshold to maximize the duration of continuous playback while minimizing the start-up delay. Liang et al. [17] establish a Markovian model to study the tradeoff between playback continuity and start-up delay. The wireless channel is modeled as a FSMC and the interplay between the channel statistics and playout buffer is provided under different buffer strategies. However, only the single-hop scenario is considered which can not be applied to multi-hop transmissions. Dua et al. [18] propose to adapt the playback threshold through a MDP. The channel is also modeled as a FSMC, where each successful transmission incurs certain profit. The playout buffer is managed to determine an optimal playback threshold to maximize the overall profit. In [19], the upper bound and lower bound of the jitter-free probability have been derived.

4.3 Analytical Framework

In this section, we first present the architecture of video streaming system. After that, we describe a general model for the playout buffer and develop an analytical framework to study the video quality perceived by the user, considering both infinite and finite buffer cases.

4.3.1 Model of Playout Buffer

Figure 4.1 shows a typical architecture of media streaming [20]. In Figure 4.1, the raw video contents are pre-compressed and saved in the storage devices. Upon the user's request, the media server retrieves the pre-stored content and then segmented it into packets. The video packets are transmitted over a lossy variable bit rate (VBR) heterogeneous network using the User Datagram Protocol (UDP)/IP protocol suite. With network dynamics, packets arrive at the user with variable delays. Without loss of generality, we assume that the inter-arrival time of video packets follows a given but arbitrary distribution with mean $\frac{1}{\lambda}$ and variance v_a. At the user end, the downloaded packets are first stored at the playout buffer, then combined into video frames and injected into the video player at the same cadence of frame rate that the video encoder generates. As the video frames are played at the constant rate, the service rate in terms of video packets is hence variable. We consider that the inter-departure time of video packets, determined by the instantaneous video playback rate, also follows a general distribution with the constant mean $\frac{1}{\mu}$ and variance v_s.

The playout buffer can accordingly be modeled as a $G/G/1/\infty$ queue when the buffer size is infinite or a $G/G/1/N$ queue when the buffer is finite. For the remaining part of this section, we analyze the evolution of the playout buffer in the infinite buffer and finite buffer cases, respectively, given network and playback statistics, i.e., λ, μ, v_a and v_s.

4.3.2 Infinite Buffer Case

We first consider the infinite playout buffer case, i.e., the buffer is infinitely large or large enough to accommodate the whole video file. This is typically true when the end host systems are personal computers with a large hard disk.

In general, the video playback process can be divided into two iterative phases, namely the charging phase and playback phase, as shown in Figure 4.2. The charging phase starts once the playout buffer becomes empty. In this case, the buffer is charged with continuously downloaded packets and the media playback is kept frozen until b packets are filled. Henceforth, we refer to b as the threshold of playback. Let R.V. (random variable) $\mathcal{D}$ denote the duration of the charging phase. The playback phase starts once the playback threshold b is reached and packets are discharged from the buffer for playback. Due to dynamic packet arrivals and departures of the buffer, the playback phase may stall when the playout buffer becomes empty again. Let R.V. $\mathcal{T}$ denote the duration of the video playback phase. The charging and playback phases iterate until the whole video is downloaded.

In this chapter, we evaluate the user's video quality in following two aspects: the *start-up delay* and *smoothness of video playback*. The former refers to the time period that users have to wait before video playback starts, which is the duration of the charging phase $\mathcal{D}$ in Figure 4.2. The latter is evaluated by the likelihood

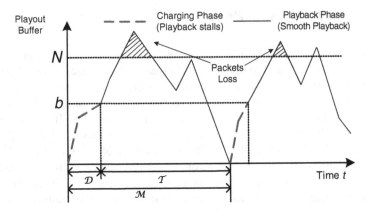

Fig. 4.2 Evolution of the playout buffer during media playout with buffer size N

or frequency of playback frozens during the media playout. The trade-off between the two aspects of video quality is adapted by the playback threshold b. A larger threshold b results in a longer start-up delay, but makes the playback less likely freeze during the media playback. In what follows, we develop a mathematical framework to investigate this trade-off and evaluate the impacts of the threshold b and network statistics on the two quality metrics.

4.3.2.1   Diffusion Approximation

To evaluate the length of start-up delay $\mathscr{D}$ and the frequency of playback frozens, we model the playout buffer as a $G/G/1/\infty$ queue and resort to the diffusion approximation [21, 22] for compact solutions [23].

Denote the buffer size at time instant t by $B(t)$. The diffusion approximation method consists in replacing the discrete buffer size $B(t)$ by a continuous process $X(t)$ and model it as the Brownian motion,

$$dX(t) = X(t+dt) - X(t) = \beta dt + G\sqrt{\alpha dt}, \qquad (4.1)$$

where $G \sim N(0,1)$ is a normally distributed random variable with zero mean and unit variance. β and α are called drift and diffusion coefficients, respectively, defined by

$$\begin{cases} \beta = E\left(\lim\limits_{\Delta t \to 0} \frac{X(t)}{\Delta t} \right) = \lambda - \mu \\ \alpha = Var\left(\lim\limits_{\Delta t \to 0} \frac{X(t)}{\Delta t} \right) = \lambda^3 v_a + \mu^3 v_s. \end{cases} \qquad (4.2)$$

Let $p(x,t|x_0)$ denote the conditional probability density function (p.d.f.) of the buffer size $X(t)$ at time t,

$$p(x,t|x_0) = \Pr\left(x \le X(t) < x + dx | X(0) = x_0 \right), \qquad (4.3)$$

where x_0 is the initial queue length. With the diffusion approximation, $p(x,t|x_0)$ can be characterized by the (forward) diffusion equation

$$\frac{\partial p(x,t|x_0)}{\partial t} = \frac{\alpha}{2}\frac{\partial^2 p(x,t|x_0)}{\partial x^2} - \beta\frac{\partial p(x,t|x_0)}{\partial x}, \tag{4.4}$$

with the initial condition

$$p(x,0|x_0) = \delta(x - x_0). \tag{4.5}$$

By applying the diffusion approximation, we can exploit the transient solution of the queue length by obtaining its p.d.f. at any time instant t.

4.3.2.2   Start-Up Delay $\mathscr{D}$

We first evaluate the start-up delay by analyzing the charging phase. In the charging phase shown in Figure 4.2, the buffer is initially empty, i.e., $x_0 = 0$, and the playback is frozen, i.e., $\mu = v_s = 0$. This phase terminates when b packets are stored. The duration of charging phase or the start-up delay is thus given by

$$\mathscr{D} = \min\{t|X(0) = 0, X(t) = b, t > 0\}. \tag{4.6}$$

Note that $\mathscr{D}$ is a random variable. In what follows, we evaluate it by showing its density function and statistics.

To this end, we model the charging phase as a diffusion process with drift $\beta_D = \lambda$ and diffusion coefficient $\alpha_D = \lambda^3 v_a$ based on (4.2). Define $P_D(x,t|0)$ as the conditional CDF of the buffer size $X(t)$ in the charging phase. During this phase, the initial buffer is empty and the queue length $X(t)$ is less than b. Thus, we have

$$P_D(x,t|0) = \Pr\{X(t) \le x|X(0) = 0, X(\tau) < b \text{ for } 0 < \tau < t\}, \tag{4.7}$$

The CDF of the start-up delay is given by

$$G_D(t) = \Pr\{\mathscr{D} \le t\} = 1 - P_D(b,t|0) = 1 - \int_0^b p_D(y,t|0)dy, \tag{4.8}$$

as $P_D(b,t|0)$ represents the probability that $X(t)$ is still below b at time t, i.e., $\mathscr{D}$ is greater than t. $p_D(x,t|0) = \Pr\{x \le X(t) < x+dx|X(0) = 0, X(\tau) < b \text{ for } 0 < \tau < t\}$ is the p.d.f. of $X(t)$ in the charging phase.

The p.d.f. of $\mathscr{D}$ is hence obtained as

$$g_D(t) = \frac{dG_D(t)}{dt} = -\frac{d}{dt}P_D(b,t|0) = -\frac{d}{dt}\int_0^b p_D(y,t|0)dy. \tag{4.9}$$

As $p_D(x,t|0)$ can be described by the diffusion approximation with the queue length never exceeding b [22], it follows the diffusion equation (4.4), as

$$\frac{\partial p_D(x,t|0)}{\partial t} = \frac{\alpha_D}{2}\frac{\partial^2 p_D(x,t|0)}{\partial x^2} - \beta_D\frac{\partial p_D(x,t|0)}{\partial x}, \quad x < b, \qquad (4.10)$$

coupled with the initial condition

$$p_D(x,0|0) = \delta(x), \qquad (4.11)$$

and the boundary condition

$$p_D(b,t|0) = 0. \qquad (4.12)$$

Equation (4.12) is obtained by the event that the diffusion process terminates when $X(t) = b$. This is imposed by the absorbing barrier in the diffusion process [24].
    Solving (4.10) with (4.11) and (4.12) yields[1]

$$p_D(x,t|0) = \frac{1}{\sqrt{2\pi\alpha_D t}}\left[\exp\left\{-\frac{(x-\beta_D t)^2}{2\alpha_D t}\right\}\right.$$
$$\left. -\exp\left\{\frac{2\beta_D b}{\alpha_D} - \frac{(x-2b-\beta_D t)^2}{2\alpha_D t}\right\}\right] \qquad (4.13)$$

and

$$P_D(x,t|0) = \Phi\left(\frac{x-\beta_D t}{\sqrt{\alpha_D t}}\right) - \exp\left\{\frac{2\beta_D b}{\alpha_D}\right\}\Phi\left(\frac{x-2b-\beta_D t}{\sqrt{\alpha_D t}}\right), \qquad (4.14)$$

where $\Phi(x) = \frac{1}{\sqrt{2\pi}}\int_{-\infty}^{x}e^{-\frac{y^2}{2}}dy$.
    Substituting (4.14) into (4.8) and (4.9), we can obtain the CDF of $\mathscr{D}$,

$$G_D(t) = 1 - \Phi\left(\frac{b-\beta_D t}{\sqrt{\alpha_D t}}\right) + \exp\left\{\frac{2\beta_D b}{\alpha_D}\right\}\Phi\left(-\frac{b+\beta_D t}{\sqrt{\alpha_D t}}\right), \qquad (4.15)$$

and its p.d.f.

$$g_D(t) = \frac{b}{\sqrt{2\pi\alpha_D t^3}}\exp\left\{-\frac{(b-\beta_D t)^2}{2\alpha_D t}\right\}. \qquad (4.16)$$

The moment generating function (m.g.f.), represented by the Laplace transform, of $g_D(t)$ is, [26]

$$g_D^*(s) = E(e^{-st}) = \exp\left[\frac{b}{\alpha_D}\left\{\beta_D - \sqrt{\beta_D^2 + 2s\alpha_D}\right\}\right]. \qquad (4.17)$$

---

[1]The solution is obtained by the method of images as shown in [24, 25].

Based on the m.g.f. $g_D^*(s)$, the mean and variance of the start-up delay with the playback threshold b can be derived accordingly,

$$E(\mathscr{D}) = -\frac{d}{ds}g_D^*(s)\bigg|_{s=0} = \frac{b}{\lambda}, \tag{4.18}$$

$$Var(\mathscr{D}) = \frac{d^2}{ds^2}g_D^*(s)\bigg|_{s=0} - E^2(\mathscr{D}) = bv_a. \tag{4.19}$$

Equations (4.18) and (4.19) indicate that the expected value and variance of start-up delay increase linearly with the playback threshold b.

4.3.2.3 Playback Duration $\mathscr{T}$

The video playback starts immediately after the charging phase. With a longer playback duration $\mathscr{T}$, less playback frozens will be encountered, and hence the length of $\mathscr{T}$ is critical to the smoothness of media playback. Without loss of generality, we focus on one playback phase and model it as a diffusion process starting at time $t = 0$. As the playback phase terminates when the buffer becomes empty again, the playback duration is thus given by

$$\mathscr{T} = \min\{t|X(0) = b, X(t) = 0, t > 0\}. \tag{4.20}$$

Denote $g_T(t)$ and $G_T(t)$ as the p.d.f. and CDF of $\mathscr{T}$, respectively. Same as the start-up delay $\mathscr{D}$, we evaluate $\mathscr{T}$ by showing its density function.

Given that the buffer size is b at the beginning of the charging phase, the probability that the buffer size $X(t)$ is larger than x at time t is given by,

$$P_T(x,t|b) = \Pr\{X(t) > x|X(0) = b, X(\tau) > 0 \text{ for } 0 < \tau < t\}$$

$$= \int_x^\infty p_T(y|t,b)dy, \tag{4.21}$$

where $p_T(y,t|b) = \Pr\{y \leq X(t) < y + dy|X(0) = b, X(\tau) > 0 \text{ for } 0 < \tau < t\}$ is the p.d.f. of $X(t)$ at time t in the playback phase, given the initial buffer size b.

Similar to the computation of start-up delay, we have

$$g_T(t) = -\frac{d}{dt}\int_0^\infty p_T(x,t|b)\,dx. \tag{4.22}$$

where $p_T(x,t)$ follows the diffusion equation,

$$\frac{1}{2}\alpha_T\frac{\partial^2 p_T(x,t|b)}{\partial x^2} - \beta_T\frac{\partial p_T(x,t|b)}{\partial x} = \frac{\partial p_T(x,t|b)}{\partial t}, \tag{4.23}$$

subject to the initial and boundary conditions

$$p_T(x,0|b) = \delta(x-b), \quad t = 0, \tag{4.24}$$

$$p_T(0,t|b) = 0, \qquad\qquad t > 0. \tag{4.25}$$

Equation (4.25) is dictated by the events that the playback phase terminates when the buffer becomes empty. β_T and α_T can be derived from (4.2).

Solving the diffusion equations (4.23)–(4.25), we have

$$p_T(x|t,b) = \frac{\exp\left\{\frac{\beta_T}{\alpha_T}(x-b) - \frac{\beta_T^2}{2\alpha_T}t\right\}}{\sqrt{2\pi\alpha_T t}}$$
$$\left[\exp\left\{-\frac{(x-b)^2}{2\alpha_T t}\right\} - \exp\left\{-\frac{(x+b)^2}{2\alpha_T t}\right\}\right]. \tag{4.26}$$

Substitute (4.26) into (4.22), we have

$$g_T(t) = \frac{b}{\sqrt{2\pi\alpha_T t^3}}\exp\left\{-\frac{(\beta_T t + b)^2}{2\alpha_T t}\right\}, \tag{4.27}$$

and its m.g.f.

$$g_T^*(s) = \exp\left\{-\frac{b}{\alpha_T}(\beta_T + \sqrt{\beta_T^2 + 2\alpha_T s})\right\}. \tag{4.28}$$

4.3.2.4   Smoothness of Playback

With the p.d.f. of $\mathcal{T}$ in hand, we are now ready to evaluate the smoothness of playback in terms of two metrics, namely the stopping probability $\mathcal{P}$ and frequency of playback frozens $\mathcal{F}$.

Stopping Probability $\mathcal{P}$: The stopping probability $\mathcal{P}$ represents the probability that the playback freezes in the middle of media playout, mathematically,

$$\mathcal{P} = \Pr(t < S|B(0) = b, B(t) = 0), \tag{4.29}$$

where S denotes the length of the video file.

Substituting (4.27) into (4.29), we have

$$\mathcal{P} = \int_0^S g_T(t)dt. \tag{4.30}$$

To obtain the closed-form expression on $\mathscr{P}$, we approximate S to be infinity as

$$\mathscr{P} \approx \lim_{S \to \infty} \int_0^S g_T(t)dt = \lim_{s \to 0} g_T^*(s),$$

$$= \begin{cases} 1, & \text{if } \beta_T \leq 0, \\ \exp\left\{-\frac{2b}{\alpha_T}\beta_T\right\}, & \text{if } \beta_T > 0. \end{cases} \tag{4.31}$$

Note that the obtained stopping probability is conservative as in reality S is limited. However, this approximation does not generate much difference as S is considerably large compared to the video frame intervals.

Substitute (4.2) into (4.31), we have

$$\mathscr{P} = \begin{cases} 1, & \text{if } \lambda \leq \mu, \\ \exp\left\{-\frac{2b}{\lambda^3 v_a + \mu^3 v_s}(\lambda - \mu)\right\}, & \text{if } \lambda > \mu. \end{cases} \tag{4.32}$$

Equation (4.32) indicates that the video playback stops with probability 1 when the mean download rate λ is less than or equal to the video playback rate μ. In the mean time, even if the mean traffic arrival rate or video download rate λ exceeds the average video playback rate μ, it is still possible that video playback stops due to the variance of packet arrivals and playback. In the real-world deployments, $\lambda - \mu$ is normally controlled small to admit more users in the system. In this case, the stopping probability $\mathscr{P}$ is heavily dependent on the threshold b and the statistics of the network.

Number of Playback Frozens $\mathscr{F}$: In (4.32), we have shown that when the mean traffic arrival rate λ is smaller than the average video playback rate μ, the video playout will stop with probability one. To shed light on how serious the interruptions of playback are in this case, we derive the overall number of playback frozens, denoted by $\mathscr{F}$, encountered during the media playback. Let R.V. $\mathscr{M}$ denote the duration between two consecutive playback frozen events, and we have $\mathscr{M} = \mathscr{D} + \mathscr{T}$, as shown in Figure 4.2. It is obvious that $\mathscr{F}$ is negatively proportional to $\mathscr{M}$. In what follows, we show its density function and statistics.

Denote the p.d.f. of $\mathscr{M}$ as $g_M(t)$. Hence,

$$g_M(t) = g_D(t) \otimes g_T(t), \tag{4.33}$$

where $\otimes$ denotes convolution. The m.g.f. of $g_M(t)$ is thus given by

$$g_M^*(s) = g_D^*(s) \cdot g_T^*(s). \tag{4.34}$$

Substitute (4.17) and (4.28) into (4.34), we can obtain the mean and variance of $\mathscr{M}$ as

$$E(\mathscr{M}) = -\frac{d}{ds}g_M^*(s)\bigg|_{s=0} = \frac{-b\mu}{\lambda(\lambda - \mu)}, \quad \lambda < \mu, \tag{4.35}$$

and

$$Var(\mathcal{M}) = \frac{d^2}{ds^2}g_M^*(s)\bigg|_{s=0} - E^2(\mathcal{M})$$

$$= -b\frac{\mu^3(v_s+v_a)+3v_a\lambda\mu\,(\lambda-\mu)}{(\lambda-\mu)^3}, \quad \lambda < \mu. \tag{4.36}$$

Next, we use the diffusion approximation to obtain the p.d.f. of $\mathcal{F}$. Specifically, we assume that there is a virtual event buffer B_F which counts the events of playback frozen. Whenever an event of playback frozen happens, we increase the queue length of B_F by one. Thus, the buffer size of B_F at time t, denoted by $X_F(t)$, represents the number of playback frozens up to time t. The interarrival time between two consecutive increments of $X_F(t)$ is $\mathcal{M}$, where $X_F(t)$ is a non-decreasing function of time t. Denote by $P_F(x,t|0)$ the conditional CDF of $X_F(t)$ at time t, given the initial buffer size 0,

$$P_F(x,t|0) = \Pr\{X_F(t) \le x | X_F(0) = 0\}. \tag{4.37}$$

Similarly, $X_F(t)$ can be approximated as a continuous function by applying diffusion equation, and its CDF is governed by

$$\frac{\partial P_F(x,t|0)}{\partial t} = \frac{\alpha_F}{2}\frac{\partial^2 P_F(x,t|0)}{\partial x^2} - \beta_F\frac{\partial P_F(x,t|0)}{\partial x}, \tag{4.38}$$

coupled with the boundary condition

$$\begin{cases} \lim_{x\to\infty} P_F(x,t|0) = 1, \, t \ge 0, \\ \lim_{x\to 0} P_F(x,t|0) = 0, \, t \ge 0. \end{cases} \tag{4.39}$$

where $\beta_F = \frac{1}{E(M)}$ and $\alpha_F = \frac{Var(M)}{E^3(M)}$ can be derived from (4.2), (4.35) and (4.36). Solving (4.38) and (4.39), we have

$$P_F(x,t|0) = \Phi\left(\frac{x - \beta_F t}{\sqrt{\alpha_F t}}\right) - \exp\left\{\frac{2\beta_F x}{\alpha_F}\right\}\Phi\left(-\frac{x + \beta_F t}{\sqrt{\alpha_F t}}\right). \tag{4.40}$$

The mean and variance of the number of playback frozens at time t, when $\lambda < \mu$, can be approximated as

$$E(\mathcal{F}) \approx \beta_F t = -\frac{\lambda\,(\lambda-\mu)}{\mu b}t, \tag{4.41}$$

$$Var(\mathcal{F}) \approx \alpha_F t = \frac{\mu^2\lambda^3(v_s+v_a)+3v_a\lambda^4\,(\lambda-\mu)}{b^2\mu^2}t, \tag{4.42}$$

as $\exp\{\frac{2\beta_F x}{\alpha_F}\}\Phi\left(-\frac{x+\beta_F t}{\sqrt{\alpha_F t}}\right)$ decreases dramatically when t is large.

4.3.3 Finite Buffer Case

In this subsection, we consider the case that the playout buffer is limited compared with the volume of video file. This is typical when the end users use personal devices with limited buffer size and hard disk such as handsets.

The start-up delay $\mathscr{D}$ obtained in the previous subsection is also valid in the finite buffer case as the start and termination conditions of the charging phase in both cases are the same. As shown in Figure 4.2, in the playback phase, the queue length of the playout buffer is upper bounded by the buffer size, denoted by N ($N > b$). When the playout buffer is full, the arrival video packets will be dropped, which not only degrades the user's video quality but also results in the bandwidth waste. Therefore, a key performance metric in this case is the packet loss probability due to buffer overflow. In this chapter, the packet loss probability and the buffer overflow probability are interchangeably used.

Let $\mathscr{L}$ denote the packet loss probability of the playout buffer,

$$\mathscr{L} = \lim_{t \to \infty} \Pr\left(B(t) = N\right). \tag{4.43}$$

The smoothness of playback is evaluated by the charging probability, denoted by $\mathscr{C}$, which is defined as the probability that the playback is frozen and the playout buffer is in the charging phase at any time instant.

We invoke the diffusion approximation to analyze playback phase in the finite buffer case and evaluate $\mathscr{L}$ and $\mathscr{C}$ in terms of network statistics and threshold of playback, as

$$\frac{\partial p(x,t|b)}{\partial t} = \frac{1}{2}\alpha_T \frac{\partial^2 p(x,t|b)}{\partial x^2} - \beta_T \frac{\partial p(x,t|b)}{\partial x}$$
$$+ \frac{\lambda}{b}\mathscr{C}\delta(x-b) + \mu\mathscr{L}\delta(x-N+1), \tag{4.44}$$

$$\lim_{x \to 0}\left[\frac{\alpha_T}{2}\frac{\partial p(x,t|b)}{\partial x} - \beta_T p(x,t|b)\right] = \frac{\lambda}{b}\mathscr{C}, \tag{4.45}$$

$$\lim_{x \to N}\left[\frac{\alpha_T}{2}\frac{\partial p(x,t|b)}{\partial x} - \beta_T p(x,t|b)\right] = -\mu\mathscr{L}, \tag{4.46}$$

subject to the initial and boundary conditions

$$\lim_{x \to 0^+} p(x,t|b) = 0 \quad t > 0,$$

$$\lim_{x \to N^-} p(x,t|b) = 0 \quad t > 0,$$

where $\delta(x)$ is the Dirac delta function; $p(x,t|b) = \Pr\{x \le X(t) < x+dx | X(0)=b\}$ is the conditional p.d.f. of the queue length $X(t)$.

The probability density in (4.44) is composed of two parts, the p.d.f. of the queue length $p(x,t|b)$ when $x \in (0,N)$ and the p.m.f. $\mathscr{L}$ and $\mathscr{C}$ on the two boundaries when buffer is full and in the charging phase, respectively. $\frac{\lambda}{b}\mathscr{C}\delta(x-b)$ in (4.44) evaluates the probability that the queue changes from the charging phase to the playout phase, where $\frac{\lambda}{b}$ is the mean rate of the change computed as $\frac{1}{E(\mathscr{D})}$. $\mu\mathscr{L}\delta(x-N+1)$ evaluates the probability that the queue length jumps from N to $N-1$ with packets being served at the mean rate μ. More details about the rational behind the equations can be found in [27].

It is important to note that the diffusion process of playback phase in (4.44) is different from that in (4.23). Equation (4.23) describes a single playback phase that starts when b packets are stored in the charging phase and terminates once the playout buffer becomes empty. However, (4.44) represents the whole media playback process from the first playback to the instant when video is wholly downloaded. The reason we consider the whole session of video playback as one diffusion process in this case is that it facilitates to evaluate the long-term buffer overflow and underflow probability in the steady state. While in the infinite buffer case, we are more interested in the transient behavior of the queue to evaluate the duration of each start-up delay and playback phase.

Solving (4.44) at the steady state when $\lim\limits_{t\to\infty} \frac{\partial p(x,t|b)}{\partial t} = 0$, we have

$$p(x,\infty|b) = \begin{cases} \frac{\lambda\mathscr{L}}{b\beta_T}(e^{rx}-1), & 0 < x \le b, \\ \frac{\lambda\mathscr{L}}{b\beta_T}(1-e^{-rb})e^{rx}, & b < x \le N-1, \\ \frac{\mu\mathscr{C}}{\beta_T}\left(1-e^{r(x-N)}\right), & N-1 < x < N, \end{cases} \tag{4.47}$$

where $r = \frac{2\beta_T}{\alpha_T}$, and the packet loss probability $\mathscr{L}$ and the charging probability $\mathscr{C}$ are given by

$$\mathscr{L} = \left(\frac{-(1-e^{-r})\mu^2 b}{\lambda\beta_T(1-e^{-rb})e^{r(N-1)}} + \frac{\lambda}{\beta_T}\right)^{-1}, \tag{4.48}$$

$$\mathscr{C} = \left(-\frac{\mu}{\beta_T} + \frac{\lambda^2}{\beta_T b\mu}\frac{e^{r(N-1)}(1-e^{-rb})}{1-e^{-r}}\right)^{-1}. \tag{4.49}$$

The infinite playout buffer could be regarded as a special case of the finite playout buffer when $N \to \infty$. In specific, when $\beta_T < 0$, i.e., $\lambda < \mu$, with (4.48) and (4.49), we have

$$\lim_{N\to\infty} \mathscr{L} = 0, \quad \lambda < \mu, \tag{4.50}$$

as no packets will be lost when $N \to \infty$, and

$$\lim_{N \to \infty} \mathscr{C} = -\frac{\beta_T}{\mu}, \quad \lambda < \mu. \tag{4.51}$$

Equation (4.51) matches the results of infinite buffer case as

$$\lim_{N \to \infty} \mathscr{C} = \frac{E(\mathscr{F}) \times E(\mathscr{D})}{t}, \quad \lambda < \mu, \tag{4.52}$$

where $E(\mathscr{D})$ and $E(\mathscr{F})$ are given in (4.18) and (4.41), respectively.

Therefore, we show the video quality in terms of start-up delay and smoothness of video playback with given statistics of the download and playback rates of video packets. The mean video playback rate μ is usually fixed and given for non-scalable video. When scalable video coding, e.g., layer-encoded video streaming [28], is used, the playback rate can be adjustable when different video layers are downloaded. The statistics of playback rate in this case are no longer predefined and therefore need to be measured at the receiver.

4.4   Quality Driven Playout Buffer Management

In this section, by exploiting the obtained video quality metrics from the analytical framework, we determine the optimal playback threshold to achieve the maximal user utility, based on different video requirements of end users. Towards this goal, we formulate the playback threshold selection as a stochastic optimization problem.

4.4.1   Infinite Buffer Case

Let $\hat{\mathscr{D}}$ and $\hat{\mathscr{F}}$ denote the maximum tolerable start-up delay and number of playback frozens input by the users, respectively. Our objective is to manage the threshold of playback b to maximize the user perceived video quality within the tolerable range specified by $\hat{\mathscr{D}}$ and $\hat{\mathscr{F}}$, mathematically,

$P1$: if $\lambda > \mu$,

$$\min_{b} \mathscr{P} + \varpi_1 \left(E(\mathscr{D}) + \vartheta_D Var(\mathscr{D}) \right)$$
$$s.t. \quad \Pr \left\{ \mathscr{D} > \hat{\mathscr{D}} \right\} \leq \zeta, \tag{4.53}$$
$$b > 0.$$

$P2 : \text{if } \lambda \leq \mu,$

$$\min_{b} \; E(\mathscr{F}) + \vartheta_F Var(\mathscr{F}) + \varpi_2 \left(E(\mathscr{D}) + \vartheta_D Var(\mathscr{D}) \right)$$

$$\text{s.t.} \qquad \Pr\left\{ \mathscr{D} > \hat{\mathscr{D}} \right\} \leq \zeta,$$

$$\Pr\left\{ \mathscr{F} > \hat{\mathscr{F}} \right\} \leq \eta, \qquad\qquad (4.54)$$

$$b > 0.$$

where $\varpi_1, \varpi_2 > 0$ are the weighting factors and $\vartheta_D, \vartheta_F \geq 0$ are called risk aversion factors which are adjustable with respect to different user requirements. ζ, η are predefined scalers such that $0 < \zeta, \eta << 1$.

Scheme $P1$ is implemented when the mean packet arrival rate λ is larger than the mean video playback rate μ. In this case, with probability $1 - \mathscr{P}$ the video playback can be finished without any interruptions. The objective is hence to avoid playback frozens while minimizing the start-up delay. ϖ_1 in the utility function is a knob to balance the requirements between smooth playback and start-up delay. A larger ϖ_1 represents that users are more sensitive to the start-up delay, e.g., when watching a live soccer match. ϑ_D is called risk aversion factor which models the user's attitude to the variance of start-up delay.[2] When ϑ_D is large, the users are conservative and require more strict start-up delay with low variance. The constraint is represented by a stochastic bound that the resulting start-up delay must be within the tolerable region $\hat{\mathscr{D}}$ imposed by the user with a high probability. The stochastic QoS is considered because providing absolute QoS guarantee may not be feasible and is typically difficult and costly for implementation in the time-varying environment [29].

The scheme $P2$ is employed when the mean packet arrival rate is insufficient to meet the playback. In this case, interruptions of playback are inevitable as shown by (4.32). The objective is to minimize the number of playback frozens and the incurred start-up delay. The utility functions and constraints are defined in the same fashion of $P1$.

Both $P1$ and $P2$ are probability-constrained stochastic optimization (also referred to as chance constrained programming) [30]. By substituting (4.15), (4.18), (4.19) and (4.32) into $P1$; (4.15), (4.18), (4.19), (4.40), (4.41) and (4.42) into $P2$, we have $P1' : \text{if } \lambda > \mu,$

$$\min_{b} \quad \exp\left\{ -\frac{2b}{\lambda^3 v_a + \mu^3 v_s}(\lambda - \mu) \right\} + \varpi_1 b \left(\frac{1}{\lambda} + \vartheta_D v_a \right)$$

$$\text{s.t.} \quad \Phi\left(\frac{b - \lambda \hat{\mathscr{D}}}{\sqrt{\lambda^3 v_a \hat{\mathscr{D}}}} \right) - \exp\left\{ \frac{2b}{\lambda^2 v_a} \right\} \Phi\left(-\frac{b + \lambda \hat{\mathscr{D}}}{\sqrt{\lambda^3 v_a \hat{\mathscr{D}}}} \right) \leq \zeta, \qquad (4.55)$$

$$b \geq 0,$$

---

[2]This utility function is defined in the fashion of Markowitz mean-variance model which is widely used in portfolio optimization.

$P2'$: if $\lambda \leq \mu$,

$$\min_{b} \quad \frac{A}{b} + \frac{\vartheta_F}{b^2}B + \varpi_2 b\left(\frac{1}{\lambda} + \vartheta_D v_a\right)$$

$$s.t. \quad \Phi\left(\frac{b-\lambda\hat{\mathscr{D}}}{\sqrt{\lambda^3 v_a\hat{\mathscr{D}}}}\right) - \exp\left\{\frac{2b}{\lambda^2 v_a}\right\}\Phi\left(-\frac{b+\lambda\hat{\mathscr{D}}}{\sqrt{\lambda^3 v_a\hat{\mathscr{D}}}}\right) \leq \zeta,$$

$$1 - \Phi\left(\frac{\hat{\mathscr{F}}-\beta_F S}{\sqrt{\alpha_F S}}\right) + \exp\left\{\frac{2\beta_F\hat{\mathscr{F}}}{\alpha_F}\right\}\Phi\left(-\frac{\hat{\mathscr{F}}+\beta_F S}{\sqrt{\alpha_F S}}\right) \leq \eta,$$

$$b \geq 0,$$

(4.56)

where $A = -\frac{\lambda(\lambda-\mu)}{\mu}S, B = \frac{\mu^2\lambda^3(v_s+v_a)+3v_a\lambda^4(\lambda-\mu)}{\mu^2}S$ are positive scalers. Here, the statistics of network and video playback rate, i.e., λ, v_a, μ and v_s, and the video length S are known and used as inputs to the control scheme. This is reasonable as those network statistics can be measured in real time at the user end.

Both $P1'$ and $P2'$ are nonlinear programming problems which may be prohibitively expensive for practical real-time streaming systems. To reduce the computation complexity, we apply the one-sided Chebyshev inequality, which states that for any R.V. χ and any positive real number x,

$$\Pr\{\chi - E(\chi) \geq x\} \leq \frac{Var(\chi)}{Var(\chi) + x^2}, \quad \text{for } \chi > E(\chi).$$

(4.57)

Using the Chebyshev inequality, together with (4.18), (4.19), (4.41) and (4.42), the constraints of $P2$ become

$$b \leq \hat{\mathscr{D}}\lambda + \frac{v_a(1-\zeta) - \sqrt{\frac{4\hat{\mathscr{D}}\zeta}{\lambda}v_a(1-\zeta) + v_a^2(1-\zeta)^2}}{2\zeta/\lambda^2},$$

(4.58)

$$b \geq \frac{A}{\hat{\mathscr{F}}} + \frac{\sqrt{B\eta(1-\eta)}}{\eta\hat{\mathscr{F}}},$$

(4.59)

where A and B are the same as those in (4.56). The details are shown in Appendix 4.7.

By replacing the constraints of $P1$ and $P2$ with (4.58) and (4.59), both $P1$ and $P2$ become convex optimization problems which could be solved efficiently. Note that computation complexity is reduced at the expanse of user's utility, because comparing with $P1'$ and $P2'$, the new constraints obtained with the Chebyshev inequality are more conservative, resulting in a smaller feasible region. However, a conservative but fast algorithm is desirable for practical use.

To ensure that the resultant video performance is within the tolerable region, the threshold of playback b must be within the range specified by (4.58) and (4.59). To make this condition satisfied, we could apply call admission control at the user end. In this way, the request of playback is reject directly at agent of the end host without sending it to the media server if there is no positive b to meet both (4.58) and (4.59). Thus, the network resources can be efficiently utilized to provision video quality for all admitted videos.

4.4.2 Finite Buffer Case

We optimize the video playback in the finite buffer case. Our objective is to control the threshold of playback b to minimize the interruptions of video playback due to buffer empty and the packet loss caused by the buffer overflow. The minimization problem, in this case, can be represented as

$$\min_b \rho_1 \mathscr{L} + \rho_2 \frac{\mathscr{C} \times S}{E(\mathscr{D})} + \varpi_1 \left(E(\mathscr{D}) + \vartheta_D Var(\mathscr{D}) \right)$$
$$s.t. \qquad \Pr\left\{ \mathscr{D} > \hat{\mathscr{D}} \right\} \leq \zeta, \qquad (4.60)$$
$$b > 0,$$

where ρ_1 and ρ_2 are the weighting factors of packet loss and charging probabilities, respectively. ϖ_1, ϑ_D and ζ are defined in the same manner as those in the infinite buffer case.

In (4.60), $\frac{\mathscr{C} \times S}{E(\mathscr{D})}$ represents the mean number of playback frozens where $\mathscr{C} \times S$ computes the overall time spent in the charging phase. The objective is to balance the trade-off between the video quality metrics, i.e., the smoothness of playback, packet loss and the encountered start-up delay.

In a summary, this section provides examples to apply the achieved analytical results to the optimal receiver buffer design. This leverages the property that the analytical results bridge the network throughput with the user perceived video quality as,

$$(\mathscr{D}, \mathscr{F}) = f(\lambda, v_a, \mu, v_s, b), \qquad (4.61)$$

where the mapping function $f(\cdot)$ could be represented by the constraints of $P1$ and $P2$ [or (4.58) and (4.59)]. In a reverse manner, we can also obtain the desired network resource with given user requirements as

$$(\lambda, v_a) = f^{-1}(\mathscr{D}, \mathscr{F}, \mu, v_s, b). \qquad (4.62)$$

This is useful as the guideline of the network resource allocation to achieve specific video quality requirements. For non-scalable video coding, the video playback rate μ is usually fixed and only the playback threshold b is adjusted to adapt to the required video quality. In this case, the presented optimization framework can be applied directly. When layered-encoded video coding is used, the video playback rate could also be adjustable [28] and the problem can be extended to a joint optimization framework of playback threshold and playback rate (or video layers) selections. We will pursue the joint optimization problem in our future work.

Table 4.1 Statistics of video frames

Video clip name	Frame number	Frame size		Bit rate		Inter-departure of pkts	
		Mean (bytes)	Variance	Mean (bit/s)	Peak	$\frac{1}{\mu}$(ms)	Variance
Aladdin	89998	7.7e+02	5.8e+05	1.5e+05	1.3e+06	33.6	102
Susi and Strolch	89998	5.8e+02	3.9e+05	1.2e+05	1.3e+06	36.2	70.4

4.5 Simulation Verification

In this section, we verify the achieved analytical results using extensive simulations, based on a trace-driven discrete event simulator coded in C++.

4.5.1 Simulation Setup

We use two real VBR video clips, "Aladdin" and "Susi and Strolch", from [31] encoded by MPEG-4 with diverse frame statistics. Each video clip lasts $S = 1$ h and the sequences are encoded at a constant frame rate of 25 frames/s in the Quarter Common Intermediate Format (QCIF) resolution (176×144). The statistics of video frames are summarized in Table 4.1.

The simulated network is shown in Figure 4.1. In each simulation run, the simulator loads video frames from the video trace file and segment the variable size video frames into IP packets with the maximum size of 1,400 Bytes. The available bandwidth of the network varies over time during which the overall variable bit rates of the channels are 10 Kbps, 500 Kbps, 2 Mbps, and 4 Mbps with probability 0.02, 0.48, 0.30 and 0.20, respectively. The average throughput is thus 1.64 Mbps; the mean and standard deviation of network delay are $\frac{1}{\lambda} = 35.4$ ms and $\sqrt{v_a} = 155.2$ ms, respectively. The video file is played at a constant rate 25 frames/s by default and variable packet rates as shown in Table 4.1. For each scenario, we conduct 30 simulation runs and plot the mean results with the 95 % confidence intervals.

4.5.2 Infinite Buffer Case

We first examine the case of infinite playout buffer.

4.5.2.1 Start-Up Delay $\mathscr{D}$ and Playback Duration $\mathscr{T}$

In the first simulation, we verify the analysis of the start-up delay $\mathscr{D}$ and playback duration $\mathscr{T}$. We use the trace "Aladdin" in which $\lambda < \mu$ according to Table 4.1. In this case, the playback frozens are inevitable as indicated by (4.32).

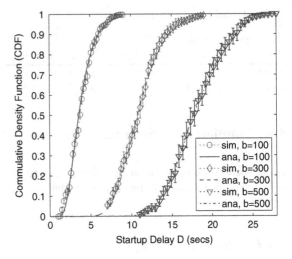

Fig. 4.3 CDF of the start-up delay D for $b = 100, 300$ and 500 packets, respectively

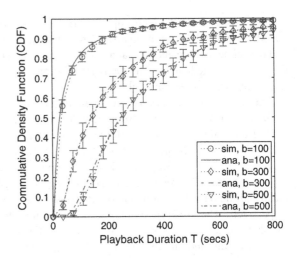

Fig. 4.4 CDF of the playback duration T for $b = 100, 300$ and 500 packets, respectively, at time $t = S$

The CDF of the start-up delays and playback durations with different buffer thresholds b are shown in Figures 4.3 and 4.4, respectively. In Figure 4.3, the mean start-up delay increases with b and the corresponding CDF moves to the left. In addition, the variance of start-up delay increases accordingly as the CDF curve expands in width. Similarly, it can be seen in Figure 4.4 that both mean and variance of the playback duration increase with the threshold b. The simulation results well validate our analysis.

Fig. 4.5 The simulated
stopping probability $\mathscr{P}$

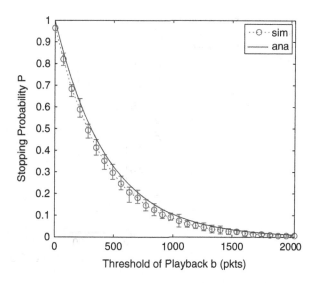

4.5.2.2   Stopping Probability $\mathscr{P}$

We verify the stopping probability $\mathscr{P}$ in the second simulation using the clip
"Susi and Strolch" where $\lambda > \mu$. In this case, the video packets are downloaded
at a faster rate than that of the playback, and the stopping probability $\mathscr{P}$ is less
than one as shown in (4.32). We conduct 30 experiments, and for each experiment
we increase the buffer threshold b by 70 packets starting from 1 packet. Within
each experiment, we conduct 500 simulation runs with each run terminated either
when the playback frozen occurs or after the whole video is played without any
interruptions. The simulation stopped by playback frozens are called frozen events.
The probability of stopping is then computed as the total number of frozen events
divided by 500. It is observed in Figure 4.5 that the probability of stopping decreases
exponentially with the increase of buffer threshold b. The analytical results are
slightly larger than the simulation results because the video length S is assumed
infinity for analysis while S is 1 h in the simulations.

4.5.2.3   Number of Playback Frozens $\mathscr{F}$

We study the number of playback frozens using the clip "Aladdin" with $\lambda < \mu$. In
this simulation, we conduct 500 runs and measure the number of playback frozens.
Figure 4.6 plots the CDF of the number of playback frozens when b is 100, 300
and 500 packets, respectively, at time $t = S$. The analysis obtained from (4.40)
well match the simulation result. Meanwhile, we can see that when b increases, the
CDF curve shifts to the left which implies that on average fewer events of playback
frozens are encountered. However, the step size of each shift is different; the mean
number of playback frozens decreases dramatically when b is initially small. The

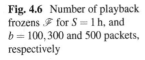

Fig. 4.6 Number of playback frozens $\mathscr{F}$ for $S = 1$ h, and $b = 100, 300$ and 500 packets, respectively

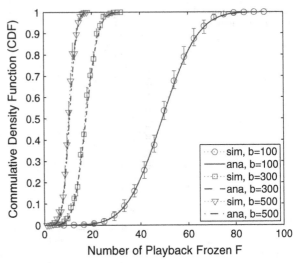

Table 4.2 Parameters in optimal playout buffer management (infinite buffer)

Scheme	Video trace	$\hat{\mathscr{D}}$	$\hat{\mathscr{F}}$	ζ	η	ϖ_1	ϖ_2	ϑ_D	ϑ_F
$P1(\lambda > \mu)$	Susi and Strolch	120 s	N/A	5%	N/A	0.01	N/A	1	N/A
$P2(\lambda \leq \mu)$	Aladdin	120 s	20	5%	5%	N/A	0.1	1	1

width of the CDF curves also becomes smaller with a larger b, which implies that the variance of the number of playback frozens decreases when b increases.

4.5.2.4   Optimal Selection of Playback Threshold b

Based on the above analytical results, we apply Matlab `fmincon` to solve (4.53) and (4.54) subject to the constraints (4.58) and (4.59) for optimal playback threshold management. The default setting of parameters are in Table 4.2.

We first show the impacts of weighting factors on the optimal selection of playback threshold. Figure 4.7 plots the optimal threshold b^* of $P1$ using trace "Aladdin". In this scenario, we can see that the optimal threshold b^* decreases monotonically with the increasing weighting factor ϖ_1. This is because when ϖ_1 in (4.53) increases, the utility of start-up delay, evaluated as $E(\mathscr{D}) + \vartheta_D Var(\mathscr{D})$, becomes more important in the objective and overwhelms the stopping probability. Therefore, the optimal threshold b^* is reduced accordingly to shrink the start-up delay at the cost of a higher playback frozen probability. The resultant stopping probability and utility of start-up delay at different optimal thresholds b^* are shown in Figure 4.8, where the utility of start-up delay is computed as $b^* \left(\frac{1}{\lambda} + \vartheta_D v_a \right)$ which is the portion of utility characterized by the start-up delay in (4.55). Figure 4.9 plots the optimal thresholds b^* of $P2$ with the increasing weighting factor ϖ_2. We can see that the corresponding b^* decreases, as the utility of start-up delay becomes

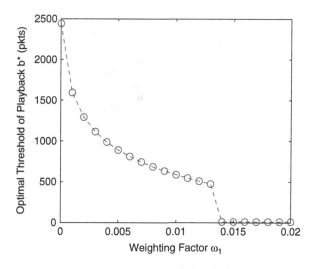

Fig. 4.7 The optimal playback threshold b^* with the increasing weighting factor ϖ_1

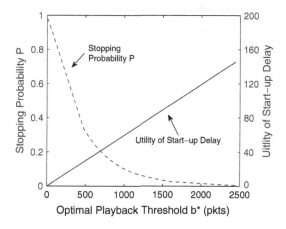

Fig. 4.8 Trade-off between the stopping probability and start-up delay at different optimal playback thresholds b^*

more critical when ϖ_2 increases. When ϖ_2 is very small, b^* is upper bounded by 2,500 packets to guarantee that the resulting start-up delay is within the tolerable value $\hat{\mathscr{D}}$. When ϖ_2 is very large, b^* is lower bounded by 800 packets to assure that the tolerable value $\hat{\mathscr{F}}$ is not violated. The resultant utilities of playback frozens and start-up delay at the different optimal thresholds of playback b^* are shown in Figure 4.10 where the utility of playback frozens is characterized by playback frozen in (4.56) computed as $\frac{A}{b^*} + \vartheta_F \frac{B}{(b^*)^2}$. The utility of start-up delay is computed in the same manner as that in Figure 4.8.

The impacts of risk aversion factors are shown in Figures 4.11–4.13. Figures 4.11 and 4.12 show the impacts of ϑ_D in schemes $P1$ and $P2$, respectively, with all the other parameters the same as in Table 4.2. With an increasing ϑ_D, the optimal playback threshold b decreases in both schemes, $P1$ and $P2$. This is because that

Fig. 4.9 The optimal playback threshold b^* with the increasing weighting factor ϖ_2

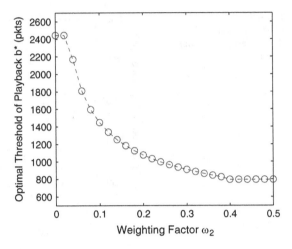

Fig. 4.10 Trade-off between the number of playback frozens and start-up delay at different optimal playback thresholds b^*

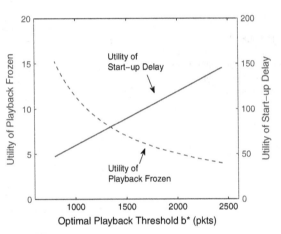

users require a strictly small variance of start-up delay. Figure 4.13 shows the impact of ϑ_F in P2. With ϑ_F increasing, the optimal playback threshold increases monotonically which hence reduce the variance of interruption frequency of the playback.

4.5.3   Finite Buffer Case

4.5.3.1   Packet Loss Probability $\mathscr{L}$ and Charging Probability $\mathscr{C}$

We verify the analytical results of the packet loss probability $\mathscr{L}$ and charging probability $\mathscr{C}$ when the buffer size is finite. In each simulation, $\mathscr{L}$ is computed as the dropped packets due to buffer overflow divided by the total number of transmitted

Fig. 4.11 Impact of θ_D on the optimal selection of playback threshold b in $P1$

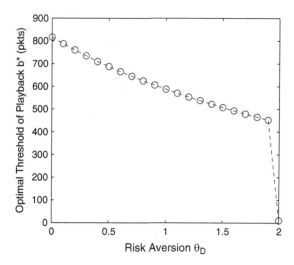

Fig. 4.12 Impact of θ_D on the optimal selection of playback threshold b in $P2$

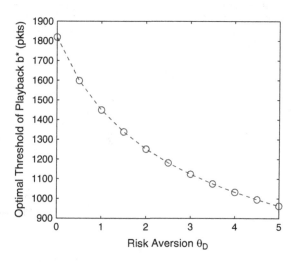

packets. $\mathscr{C}$ is evaluated as the overall time spent in the charging phase divided by the whole video session length, i.e., 1 h. We set the buffer size N to be 500 packets by default and use the trace "Aladdin" in the experiment.

Figure 4.14 plots the packet loss probability $\mathscr{L}$ with increasing threshold b. With a larger b, more packets are buffered in the charging phase and therefore the buffer becomes more easily to get filled. As a result, $\mathscr{L}$ increases monotonically with b. Figure 4.15 plots the charging probability $\mathscr{C}$ under various thresholds b. It can be seen that $\mathscr{C}$ also increases monotonically with b. This is because that the charging probability $\mathscr{C}$ represents the probability that at any point the buffer is in the charging phase; with a larger b, each charging phase is elongated, making $\mathscr{C}$

Fig. 4.13 Impact of θ_F on the optimal selection of playback threshold b in $P2$

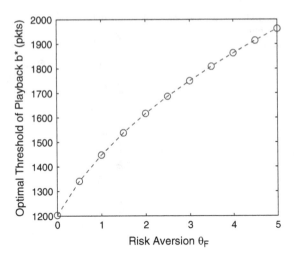

Fig. 4.14 Packet loss probability with the increasing playback threshold b

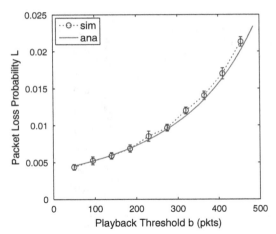

increase accordingly. However, the mean number of playback frozens, computed as $\frac{\mathscr{C} \times S}{E(\mathscr{D})}$, reduces when b increases, as shown in Figure 4.16.

Figure 4.17 plots the packet loss probability $\mathscr{L}$ and charging probability $\mathscr{C}$ under different video frame rates (and playback rate). It can be seen that, with an increasing playback rate μ which implies more video frames are played in a unit time, $\mathscr{L}$ decreases and $\mathscr{C}$ increases. This is because with a faster playback, the buffer is more likely to become empty and less likely to overflow. The simulation verifies our analysis with various values of μ. Figure 4.18 shows the impacts of the playout buffer size N on the overflow and charging probabilities when playback threshold b is 50 packets. It can be seen that as the buffer size increases, both $\mathscr{L}$ and $\mathscr{C}$ decrease monotonically. This is because that with enhanced buffer capacity,

Fig. 4.15 Charging probability with the increasing playback threshold b

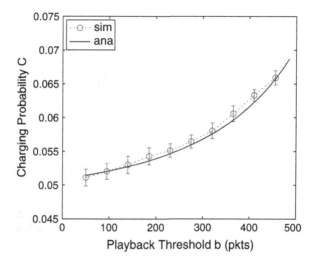

Fig. 4.16 Mean number of playback frozens with the increasing playback threshold b

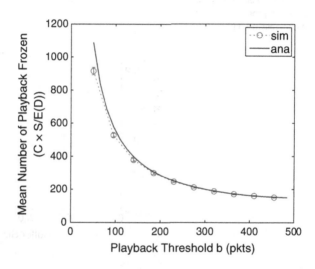

less packets will be dropped due to buffer overflow and more packets are served for playback. This reduces the frequency that buffer becomes empty. However, unlike $\mathscr{L}$ which becomes 0 when N is large, $\mathscr{C}$ approaches to a non-zero value as $\lim_{N \to \infty} \mathscr{C} = -\frac{\lambda - \mu}{\mu} = 0.0536$, as derived in (4.50) and (4.51).

4.5.3.2 Optimal Selection of Playback Threshold b

After verifying the correctness of our analysis, we show how to invoke the analytical results to optimally determine the optimal playback threshold b^* using numerical

Fig. 4.17 Packet loss and
charging probabilities with
the increasing frame rate

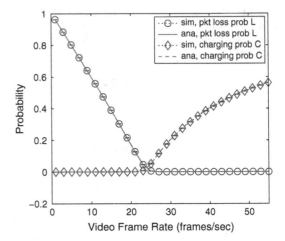

Fig. 4.17 Packet loss and
charging probabilities with
the increasing frame rate

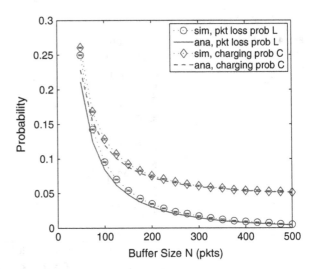

Fig. 4.18 Packet loss and
charging probabilities with
the increasing buffer size

examples. We solve (4.60) using the fmincon function of Matlab. The video
trace used is "Susi and Strolch" and the default setting of the parameters are:
$\hat{\mathscr{D}} = 15\,s, \rho_1 = 50, \rho_2 = 0.05, \varpi_1 = 0.1, \vartheta_D = 1$ and $\zeta = 0.05$.

Figure 4.19 shows the optimal playback threshold b^* with different values of ρ_1
and default setting of other parameters. It can be seen that the increasing ρ_1 leads
to the decrease of b^* because a smaller playback threshold is preferred in order
to avoid buffer overflow. Figure 4.20 shows the impact of ρ_2. When ρ_2 increases,
the objective function is sensitive to the playback frozen and hence a larger b is

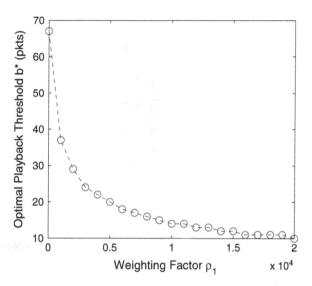

Fig. 4.19 The optimal playback threshold b^* with the increasing weighting factor ρ_1

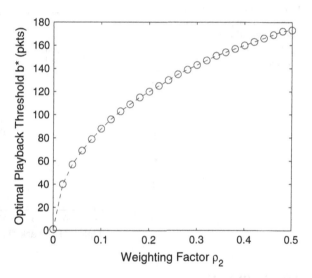

Fig. 4.20 The optimal playback threshold b^* with the increasing weighting factor ρ_2

desirable. Figure 4.21 plots the optimal selection of playback threshold b with the increasing ϖ_1. Similar to the case of infinite buffer, with ϖ_1 increasing, the optimal playback threshold b^* decreases monotonically to keep reducing the start-up delay. Figure 4.22 shows the tradeoff between the mean number of playback frozens computed by $\frac{\mathscr{C} \times S}{E(\mathscr{D})}$ and the packet loss probability $\mathscr{L}$.

Fig. 4.21 The optimal playback threshold b^* with the increasing weighting factor ϖ_1

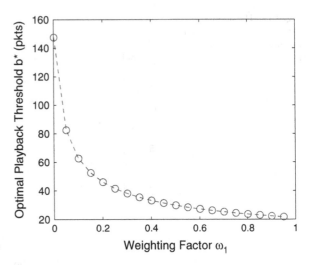

Fig. 4.22 Trade-off between the mean number of playback frozens and the packet loss probability at different optimal playback thresholds b^*

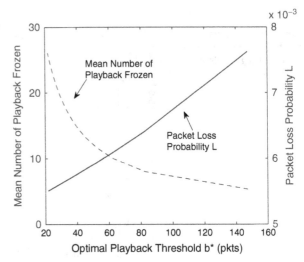

4.6   Summary

We have described a mathematical framework to study the impacts of network dynamics on the perceived video quality of end users. The presented proposal evaluates the user's perceptual quality, in terms of the start-up delay, playback smoothness, and the packet loss probability, and represents them by the network statistics and the threshold of playback in closed-form expressions. The chapter also shows how to invoke the analytical framework to guide the playout buffer design, and lastly presents simulation results to verify the analytical results.

In this chapter, we mainly focus on the user's perfective and discuss on the adaption of playback strategy at the user end according to the measurement of VBR channels. Based on the user's requirement on the playback quality, using the derived results, it is also possible to evaluate the desired network performance in terms of throughput and delay variations, and use this as an input for the network optimization [32].

For instance, [33] considers the drive-thru Internet scenario where a vehicle drives through the roadside infrastructure and subscribes to download media flows for live media presentation. Due to the volatile wireless channel, coupled with the interference and contentions among vehicles in proximity, packet delivery to vehicles in the drive-thru Internet scenario may suffer from severe packet losses. Retransmissions of packets are used to correct the errors. This, however, prolongs the packet delivery and may lead to the underflow of playout buffer, resulting in the frozen of video playback. Asefi et al. [33] models the impact of packet retransmissions on the video download rate to each user and the resultant QoE accordingly based on (4.27) and (4.32). The optimal retransmission scheme is then developed given the QoE requirement of users.

4.7 Appendix

We show how to apply the Chebyshev inequality (4.57) to derive (4.58) and (4.59), respectively.

Based on (4.57), we have

$$
\Pr\left\{\mathscr{D} > \hat{\mathscr{D}}\right\} \leq \frac{Var(\mathscr{D})}{Var(\mathscr{D}) + \left(\hat{\mathscr{D}} - E\left(\mathscr{D}\right)\right)^2}
$$
$$
= \frac{bv_a}{bv_a + \left(\hat{\mathscr{D}} - \frac{b}{\lambda}\right)^2}.
$$
(4.63)

To satisfy the constraint $\Pr\left\{\mathscr{D} > \hat{\mathscr{D}}\right\} \leq \zeta$, we make

$$
\frac{bv_a}{bv_a + \left(\hat{\mathscr{D}} - \frac{b}{\lambda}\right)^2} \leq \zeta,
$$
(4.64)

which implies

$$b \leq \hat{\mathscr{D}}\lambda + \frac{v_a(1-\zeta) - \sqrt{\frac{2D\zeta}{\lambda}v_a(1-\zeta) + v_a^2(1-\zeta)^2}}{2\zeta/\lambda^2} \quad \text{or}$$

$$b \geq \hat{\mathscr{D}}\lambda + \frac{v_a(1-\zeta) + \sqrt{\frac{2D\zeta}{\lambda}v_a(1-\zeta) + v_a^2(1-\zeta)^2}}{2\zeta/\lambda^2}. \tag{4.65}$$

As $\hat{\mathscr{D}} \geq E(\mathscr{D}) = \frac{b}{\lambda}$, we have $b \leq \hat{\mathscr{D}}\lambda$. Together with (4.65), we have

$$b \leq \hat{\mathscr{D}}\lambda + \frac{v_a(1-\zeta) - \sqrt{\frac{2\hat{\mathscr{D}}\zeta}{\lambda}v_a(1-\zeta) + v_a^2(1-\zeta)^2}}{2\zeta/\lambda^2}. \tag{4.66}$$

Apply the Chebyshev inequality to bound $\Pr\left\{\mathscr{F} > \hat{\mathscr{F}}\right\}$, we have

$$\Pr\left\{\mathscr{F} > \hat{\mathscr{F}}\right\} \leq \frac{Var(F)}{Var(\mathscr{F}) + \left(\hat{\mathscr{F}} - E(\mathscr{F})\right)^2} = \frac{\frac{B}{b^2}}{\frac{B}{b^2} + \left(\hat{\mathscr{F}} - \frac{A}{\lambda}\right)^2}. \tag{4.67}$$

To satisfy the constraint $\Pr\left\{\mathscr{F} > \hat{\mathscr{F}}\right\} \leq \eta$, we make

$$\frac{\frac{B}{b^2}}{\frac{B}{b^2} + \left(\hat{\mathscr{F}} - \frac{A}{\lambda}\right)^2} \leq \eta, \tag{4.68}$$

which implies

$$b \leq \frac{A}{\hat{\mathscr{F}}} - \frac{\sqrt{B\eta(1-\eta)}}{\eta\hat{\mathscr{F}}} \quad \text{or} \quad b \geq \frac{A}{\hat{\mathscr{F}}} + \frac{\sqrt{B\eta(1-\eta)}}{\eta\hat{\mathscr{F}}}. \tag{4.69}$$

In addition, b be set such that $\hat{\mathscr{F}} \geq E(\mathscr{F}) = \frac{A}{b}$, i.e., $b \geq \frac{A}{\hat{\mathscr{F}}}$. Substitute it into (4.69), we have

$$b \geq \frac{A}{\hat{\mathscr{F}}} + \frac{\sqrt{B\eta(1-\eta)}}{\eta\hat{\mathscr{F}}} \tag{4.70}$$

References

1. X. Hei, C. Liang, J. Liang, Y. Liu, and K. W. Ross, "A measurement Study of a Large-scale P2P IPTV System," *Multimedia, IEEE Transactions on*, vol. 9, no. 8, pp. 1672–1687, 2007.
2. "Sandvine Report: Netflix And YouTube Account For 50% Of All North American Fixed Network Data." https://www.sandvine.com/pr/2013/11/11/sandvine-report-netflix-and-youtube-account-for-50-of-all-north-american-fixed-network-data.html, 2013.
3. B. Girod, J. Chakareski, M. Kalman, Y. J. Liang, E. Setton, and R. Zhang, "Advances in Network-Adaptive Video Streaming," *Wireless Communications and Mobile Computing*, vol. 2, no. 6, pp. 549–552, 2002.
4. Q. Zhang, W. Zhu, and Y.-Q. Zhang, "End-to-End QoS for Video Delivery over Wireless Internet," *Proceedings of the IEEE*, vol. 93, pp. 123–134, Jan. 2005.
5. P. A. Chou and Z. Miao, "Rate-Distortion Optimized Streaming of Packetized Media," *IEEE Trans. on Multimedia*, vol. 8, pp. 390–404, Apr. 2006.
6. J. Chakareski and P. Frossard, "Rate-Distortion Optimized Distributed Packet Scheduling of Multiple Video Streams over Shared Communication Resources," *IEEE Trans. on Multimedia*, vol. 8, pp. 207–218, Apr. 2006.
7. S. Mao, Y. T. Hou, X. Cheng, H. D. Sherali, S. F. Midkiff, and Y.-Q. Zhang, "On Routing for Multiple Description Video over Wireless Ad Hoc Networks," *IEEE Trans. on Multimedia*, vol. 8, pp. 1063–1074, Oct. 2006.
8. X. Tong, Y. Andreopoulos, and M. van der Schaar, "Distortion-Driven Video Streaming over Multihop Wireless Networks with Path Diversity," *IEEE Trans. on Mobile Computing*, vol. 6, pp. 1343–1356, Dec. 2007.
9. J. Xu, X. Shen, J. W. Mark, and J. Cai, "Adaptive Transmission of Multi-Layered Video over Wireless Fading Channels," *IEEE Trans. on Wireless Communications*, vol. 6, p. 2305, Jun. 2007.
10. R. A. Berry and R. G. Gallager, "Communication over Fading Channels with Delay Constraints," *IEEE Trans. on Information Theory*, vol. 48, pp. 1135–1149, May 2002.
11. Y. Eisenberg, C. E. Luna, T. N. Pappas, R. Berry, and A. K. Katsaggelos, "Joint Source Coding and Transmission Power Management for Energyefficient Wireless Video Communications," *IEEE Trans. on Circuits and Systems for Video Technology,*, vol. 12, pp. 411–424, Jun. 2002.
12. N. Laoutaris and I. Stavrakakis, "Intrastream Synchronization for Continuous Media Streams: a Survey of Playout Schedulers," *IEEE Network*, vol. 16, pp. 30–40, May-Jun. 2002.
13. J. Liu, B. Li, and Y.-Q. Zhang, "An End-to-End Adaptation Protocol for Layered Video Multicast using Optimal Rate Allocation," *IEEE Trans. on Multimedia*, vol. 6, pp. 87–102, Feb. 2004.
14. L. Galluccio, G. Morabito, and G. Schembra, "Transmission of Adaptive MPEG Video over Time-Varying Wireless Channels: Modeling and Performance Evaluation," *IEEE Trans. on Wireless Communications*, vol. 4, pp. 2777–2788, Nov. 2005.
15. M. Kalman, E. Steinbach, and B. Girod, "Adaptive Media Playout for Low-Delay Video Streaming over Error-Prone Channels," *IEEE Trans. on Circuits and Systems for Video Technology*, vol. 14, pp. 841–851, Jun. 2004.
16. N. Laoutaris, B. V. Houdt, and I. Stavrakakis, "Optimization of a Packet Video Receiver under Different Levels of Delay Jitter: an Analytical Approach," *Performance Evaluation*, vol. 55, no. 3–4, pp. 251–275, 2004.
17. G. Liang and B. Liang, "Balancing Interruption Frequency and Buffering Penalties in VBR Video Streaming," in *Proc. of IEEE Infocom*, 2007.
18. A. Dua and N. Bambos, "Buffer Management for Wireless Media Streaming," in *Proc. of IEEE GLOBECOM*, 2007.
19. G. Liang and B. Liang, "Effect of Delay and Buffering on Jitter-Free Streaming over Random VBR Channels," *IEEE Trans. on Multimedia*, vol. 10, Oct. 2008.

20. D. Wu, Y. T. Hou, W. Zhu, Y. Q. Zhang, and J. M. Peha, "Streaming Video Over the Internet: Approaches and Directions," *IEEE Transactions on Circuits and Systems for Video Technology*, vol. 11, pp. 282–300, Mar. 2001.

21. L. Kleinrock, *Queueing Systems, Volume II: Computer Applications*. John Wiley & Sons, 1976.

22. G. Louchard and G. Latouche, *Probability Theory and Computer Science*. Academic Press Professional, Inc. San Diego, CA, USA, 1983.

23. A. Duda, "Transient Diffusion Approximation for Some Queueing Systems," in *Proc. of ACM Sigmetrics*, 1983.

24. D. R. Cox and H. D. Miller, *The Theory of Stochastic Processes*. Chapman & Hall/CRC, 1977.

25. F. P. T. Czachorski, "Diffusion Approximation as a Modelling Tool in Congestion Control and Performance Evaluation," in *Proc. of HET-NET*, 2004.

26. M. Abramowitz and I. A. Stegun, eds., *Handbook of Mathematical Functions*. New York: Dover, 1965.

27. E. Gelenbe, "On Approximate Computer System Models," *Journal of the ACM (JACM)*, vol. 22, no. 2, pp. 261–269, 1975.

28. D. Wu, Y. T. Hou, and Y.-Q. Zhang, "Scalable Video Coding and Transport over Broadband Wireless Networks," *Proceedings of the IEEE*, vol. 89, pp. 6–20, Jan. 2001.

29. W. Kumwilaisak, Y. T. Hou, Q. Zhang, W. Zhu, C.-C. J. Kuo, and Y.-Q. Zhang, "A Cross-Layer Quality-of-Service Mapping Architecture for Video Delivery in Wireless Networks," *IEEE Journal on Selected Areas in Communications*, vol. 21, pp. 1685–1698, Dec. 2003.

30. J. R. Birge and F. Louveaux, *Introduction to Stochastic Programming*. Springer, 1997.

31. F. H. P. Fitzek and M. Reisslein, "MPEG-4 and H. 263 Video Traces for Network Performance Evaluation," *IEEE Network*, vol. 15, pp. 40–54, Nov.-Dec. 2001.

32. T. H. Luan, S. Li, M. Asefi, and X. Shen, "Quality of Experience Oriented Video Streaming in Challenged Wireless Networks: Analysis, Protocol Design and Case Study," *IEEE Multimedia communications Technical Committee E-Letter*, vol. 13, no. 1, pp. 127–139, 2012.

33. M. Asefi, J. W. Mark, and X. Shen, "A Mobility-Aware and Quality-Driven Retransmission Limit Adaptation Scheme for Video Streaming over VANETs," *IEEE Transactions on Wireless Communications*, vol. 11, no. 5, pp. 1817–1827, 2012.

Chapter 5
Conclusion

In this chapter, we summarize the main concepts and results presented in this monograph and highlight future research directions.

5.1 Concluding Remarks

In this monograph, we have investigated on the content distribution issue in vehicular networks. Based on the analysis and discussion provided throughout this monograph, we present the following remarks.

- Live and on-demand content distribution services represent the killer Internet applications and have already achieved great popularity from the Internet user's community. To deliver high-quality content distribution services to in-motion vehicles can greatly enhance the comfort of our road trips. This, however, is challenged by the instinct dynamic nature of vehicular communications due to the fast node mobility. Note that different content distribution services may have specific and heterogenous application requirements due to the nature of contents. In addition, vehicular networks in different deployment environments, like rural and urban, would also present distinguished features on vehicle mobility and available access methods. The design of content distribution applications in vehicular network should not only take the specific application requirements into accounts, but also fully explore the feature of vehicles in the specific deployment environments.

- We argue that an infrastructure which is dedicated to vehicular communications with reserved communication and storage capacities is necessary to provide reliable and QoS guaranteed vehicular content distribution services. Towards this end, there remain two issues. The first issue is on how to construct and maintain a practical infrastructure which is profitable to not only individual service users but also the vendors. Along this direction, how to incorporate the infrastructure with the existing access networks is also an important issue. The second issue

T.H. Luan et al., *Enabling Content Distribution in Vehicular Ad Hoc Networks*,
SpringerBriefs in Computer Science, DOI 10.1007/978-1-4939-0691-8_5,
© The Author(s) 2014

is how to ensure that the infrastructure works in an efficient manner so as to completely utilize the network resource and provide the best service quality by fully exploring the features of both vehicular users and the applications they subscribe.

- The rich capacity of V2V communications is undoubtedly a great treasure to enable the high-quality vehicular content distribution services, especially when infrastructure bandwidth is not available or very expensive. However, the V2V communications are by nature unreliable which conflicts with the basic integrity requirement of content distribution services. In this monograph, we have described some preliminary results to enable the integrity-oriented content transmissions over the V2V communications. How to make the proposal more practical with more accurate channel modeling deserves further study.

5.2   Future Research Directions

The efficient content distribution to vehicles includes many fundamental research challenging issues not only from the networking part but also from the service development aspect. We close this chapter and monograph with additional three thoughts on future research directions in this field.

- *Hybrid Access Networks*: As indicated in [1], 89.1% of the total population in Australia and 82.3% of that in U.S. living in city. To provide high-quality vehicular communication services in general, and content distribution in specific, in urban cities is therefore important. With the advances of wireless access technologies and especially driven by the flourish the mobile electronics, a variety of access networks like 3G/4G cellar networks, WiFi hotspots and WiMax, are now available in most cities to provide the ubiquitous high-rate connectivity in the city area. However, existing access networks have distinguished features. Moreover, they may be already overwhelmed to serve mobile users with handheld devices [2] and are not open to vehicular users. As such, how to integrate the different access methods and optimize the download strategy for vehicular users that can fully explore the vehicle's mobility feature [3] and optimize the usage of different access methods [4] deserves further study.
- *Security Issues and Incentive Mechanism Design*: With the flood of various contents in the network, to guarantee the secure download for vehicular users is the foremost issue for the vehicular content distribution services to be widely accepted [5]. This however represents several fundamental challenges which mainly attribute to the dynamic and distributed nature of the system. In specific, content distribution in vehicular networks may heavily relies on the cooperative transmission among peer vehicles. In this case, contents with virus or spyware may be dispersed by several malicious nodes. A central coordinator to audit the content transmission behaviors of vehicles may be impractical due to the large-scale network size and deployment environment like in rural and forested

highways. As such, a fully distributed mechanism, which can quickly and efficiently identify the malicious nodes and filter the contaminated contents [6], is desirable. In addition, note that the content distribution network would rely on distributed vehicle nodes to contribute their own upload bandwidth for transmitting the cached contents to others. In this case, it is important to develop a distributed and efficient incentive mechanism to spur the cooperative content distribution and punish the free-riders.

- *Service Development*: Vehicles, due to their locations and target destinations, often have prominent features such as social features and mobility features [7, 8]. The original content distribution services developed for Internet and mobile phone users, however, do not explore those features. In this case, it is interesting to design the vehicular user-oriented location- and trip-related content distribution services. Potential applications include the social content distributions and road scenario reports, etc.

References

1. http://www.theguardian.com/, "Percentage of Global Population Living in Cities, by Continent." http://www.theguardian.com/news/datablog/2009/aug/18/percentage-population-living-cities, Accessed in 2013.
2. "SXSW iPhone Users Overwhelm AT&T's 3G Coverage."
3. N. Lu, N. Zhang, N. Cheng, X. Shen, J. M. Mark, and F. Bai, "Vehicles Meet Infrastructure: Towards Capacity-Cost Tradeoffs for Vehicular Access Networks,"
4. N. Cheng, N. Lu, N. Zhang, X. Shen, and J. W. Mark, "Vehicular WiFi Offloading: Challenges and Solutions," *Vehicular Communications (Elsevier)*. URL: https://ece.uwaterloo.ca/~n7lu/jrnl_13_VCOM_CLZSM.pdf
5. H. Zhu, R. Lu, X. Shen, and X. Lin, "Security in Service-Oriented Vehicular Networks," *IEEE Wireless Communications*, vol. 16, no. 4, pp. 16–22, 2009.
6. R. Lu, X. Lin, T. H. Luan, X. Liang, X. Li, L. Chen, and X. Shen, "PReFilter: An Efficient Privacy-preserving Relay Filtering Scheme for Delay Tolerant Networks," in *Proc. of IEEE Infocom*, 2012.
7. N. Lu, T. H. Luan, M. Wang, X. Shen, and F. Bai, "Bounds of Asymptotic Performance Limits of Social-Proximity Vehicular Networks," *IEEE/ACM Trans. on Networking*. DOI 10.1109/TNET.2013.2260558.
8. R. Lu, X. Lin, and X. Shen, "SPRING: a Social-based Privacy-preserving Packet Forwarding Protocol for Vehicular Delay Tolerant Networks," in *Proc. of IEEE Infocom*, 2010.